Molluscs

J. E. Morton

D.Sc.(Lond.)

Professor of Zoology,
Auckland University, New Zealand

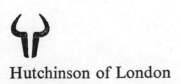

Hutchinson of London

Hutchinson & Co. (Publishers) Ltd
3 Fitzroy Square, London w1p 6jd

London Melbourne Sydney Auckland
Wellington Johannesburg and agencies
throughout the world

First published 1958
4th edition 1967
5th edition 1979

Printed in Great Britain at The Anchor Press Ltd
and bound by Wm Brendon & Son Ltd
both of Tiptree, Essex

ISBN 0 09 134160 4 cased
        0 09 134161 2 paper

To E.F.M. and R.B.M.
in gratam memoriam

Contents

Preface to fourth edition

In the eight years since first publication of this book, important writings upon molluscs have appeared. An imposing landmark is Fretter and Graham's distinguished monograph of British Prosobranchs, dealing too with some wider aspects of the phylum generally. *The Physiology of Mollusca*, edited by Wilbur and Yonge, contains valuable modern appraisals of function; and the first molluscan volume of the handsome *Traité de Zoologie* was published in 1960. M. J. Wells' short but important book on *Brain and Behaviour in Cephalopods* provides also a fascinating account of cephalopods generally. The Proceedings of the Malacological Society of London have enlarged their format and scope and several younger periodicals on the Mollusca have come into being. There is also an increasing range of regional shell-books, many of them finely illustrated. Deserving of special mention is R. Tucker Abbott's pocket-guide *Sea Shells of the World*, for its size a prodigy of information and illustration.

In the midst of all these it is pleasing to feel this book on 'Molluscs' still finds use as an introduction and connecting guide among the larger works. With this edition there has been an opportunity for a fuller revision of the text, and the correction of slips pointed out by others. The publishers have also allowed an increase in the illustrations which reviewers almost without exception found too scanty in the first editions.

Auckland   J.E.M.
October 1966

Preface to fifth edition

More students of the Mollusca have been at work in the 1970s than in any previous decade. Almost all the areas mapped out in former editions of this book have been further enlarged, by more exact and comprehensive treatment.

The great advances have been where new techniques have reopened old fields. Thus, with electronic recording, E. D. Trueman and his school have given a new impetus to locomotor studies. Isotope tracers have pushed forward our understanding of algal symbiosis, in giant clams and sacoglossans. Most significant advance of all, the electron microscope – both scanning and transmission – has reopened the whole field of fine structure. New methods of histo- and cyto-chemistry have made histology a frontier discipline again. Owen's studies of the digestive gland are a notable instance, in the wider context of the rhythmical nature of bivalve feeding and digestion.

In the classical field of origins phylogeny, where *Neopilina* has until recently held the stage, Salvini-Plawen's studies have given a new importance to the early Aplacophora. More than any other phylum, perhaps, the Mollusca have continued to be appreciated holistically with an integrated understanding of the fuctioning animal; and this at a time when 'organismic' biology has so often lost ground, on the large scale to the ecosystem and on the minute scale to the cell.

Among the more important books of the past decade have been R. D. Purchon's *Biology of the Molluscs*, Libbie Hyman's first (and regrettably only) molluscan volume in *The Invertebrates*; two collected volumes from the Zoological Society Symposium on Molluscs (1968), and *Pulmonates*, edited by Vera Fretter and J. F. Peake. Runham and Hunter have also produced a useful volume on pulmonate slugs. Yonge and Thompson's well illustrated *Living Marine Molluscs* appeared in 1976. This revision was completed before full justice could be done to T. E. Thompson's first Ray Society volume (1976) on Opisthobranchia. The volume from the

International Malacological Conference in 1973 is also an important source-book.

Of the three leaders of British malacology saluted in the preface to the first edition (1958), it is inspiring to see Sir Maurice Yonge now well entered upon his second half-century of active research; while from Vera Fretter and Alastair Graham came in 1976 *A Functional Anatomy of Invertebrates*, with an elegant and compact chapter on the living mollusc.

Fine books are every year getting more expensive. This must be the chief justification for the reissue of a small one, giving an introductory perspective, with some guidance through the literature, and to the abounding diversity of the animals themselves.

1 Introduction and general features

It is not difficult to recognize a mollusc. One looks first for a hard shell, a soft body and a slippery skin. The usual term 'shellfish' is too narrow, however, to include slugs and squids, and there is indeed no English vernacular name taking in the whole phylum Mollusca. In essentials molluscs are one of the most compact groups of animals; but there can be few phyla that show such wide diversity imposed on such a uniform plan. Most molluscs have no internal skeleton, and no stereotyped pattern, as in a segmented worm or a jointed arthropod. There is no standard molluscan shape, and in an evolutionary sense molluscs are plastic material. The outlines of the body are freely altered as new habits are acquired and new structures are needed. Though there are many exceptions, most molluscs are slow-moving and confined to rather special habitats. They bear the adaptive stamp of the environment in a far more obvious way than more active animals that can move about widely.

Molluscs range from limpets clinging to the rocks, to snails which crawl or dig or swim, to bivalves which anchor or burrow or bore, to cephalopods which torpedo through the water or lurk watchfully on the bottom. They penetrate all habitats: the abysses of the sea, coral reefs, mudflats, deserts and forests, rivers, lakes and under ground. They may become hidden as parasites in the interior of other animals. They feed on every possible food and vary in size from giant squids and clams to little snails a millimetre long. In number of species, the Mollusca are the second phylum to the Arthropoda: it is almost impossible to assess accurately the total of species – probably about 80000, as compared with eight times as many insects but only half as many vertebrates. Three-quarters of the molluscs are gastropods, with about 1650 genera. The lamellibranchs come next with 420. Third in numbers, but greatest in size, are the cephalopods, with 150 genera.

There are six classes of molluscs, and it is hard for any definition to take account of all their variations. In this book we shall survey the Mollusca from an evolutionary and adaptive point of view, and try to recognize the trends running through the history of each class. As others have done before, we must first present our own concept of the early, primitive mollusc. The idea of an 'archi-mollusc' is a favourite and well-worked one. We have already had archetypal snails, with conical shells poised at rakish angles, and strange pelagic molluscs – half nautilus and half veliger. The danger is that in mixing genealogical ideas with morphology our archetype may become like an heraldic animal – a lowest common multiple of incompatible organs.

A possible early mollusc has been reconstructed in Fig. 1. It is a slow-moving animal, with a low conical shell and a broad creeping surface. It probably lived on a firm substratum in shallow inshore waters. The molluscan body has two well-marked regions. The anterior and lower part is firm and muscular and normally lies outside the shell, though it can be wholly retracted into it. This part is sometimes known as the 'head-foot'. Though it has no common name it is a real structural entity in all molluscs. It primitively forms a flat foot on which the animal crawls by muscular waves, and a head carrying a short snout bearing the mouth, and with eyes and tentacles. The second part of the body is dorsal and posterior, and never leaves the shell. It forms a soft, thin-walled visceral mass, entirely non-muscular. Attached to the visceral mass, and hanging freely from it, it is a wide skirt called the mantle or pallium. This lines the whole of the shell, which is secreted by its interior and free margin. A large space lies between the mantle and the sides of the body. It is deepest behind, where it is known as the *mantle cavity* or *pallial cavity*, morphologically 'outside' the body, and forming – like the shell – one of the leading features of the Mollusca.

The head-foot, though covered with cilia and mucous cells, works chiefly by the action of muscles. The visceral mass and the pallial cavity at this early stage rely most on the action of mucus and cilia. The mollusc is thus divided into separate muscular and ciliary components. The muscular animal is responsible for locomotion, retreat into the shell and capture of food; the ciliary animal carries out almost all the other functions of the gut and viscera, and of the organs in the pallial cavity.

The most prominent mantle organs are a pair of gills or *ctenidia*, one lying at either side of the cavity. Each is made up of two rows of triangular leaflets or *filaments* lying at either side of a central axis. Quite complicated tracts of cilia are developed upon the gill filaments.

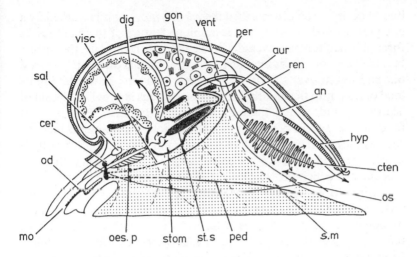

Figure 1 Schematic view of an early mollusc
an, anus; aur, auricle; cer, cerebral ganglion; cten, ctenidium; dig, digestive gland; gon, gonad; hyp, hypobranchial gland; mo, mouth; od, odontophore; oes.p, oesophageal pouches; os, osphradium; ped, pedal cord; per, pericardium; ren, renal organ; sal, salivary gland; s.m, shell muscle; stom, stomach; st.s, style sac; vent, ventricle; visc, visceral loop

Some of these draw a respiratory water current into the mantle cavity, others intercept unwanted particles that the current carries in. A water current enters the pallial cavity at either side and ventrally to each gill. It is tested before it reaches the gill by a pallial sense organ – the *osphradium*. Filtering between the gill filaments, the currents meet near the dorsal mid-line, where an exhalant stream passes backwards out of the pallial cavity. In the line of exit lies the anus, and at either side of it is a renal pore, from which passes nitrogenous waste, and genital products as well. The sanitation of the pallial cavity is an early preoccupation of molluscs. Rejectory cilia on the gill filaments throw off waste towards the mid-line. As well as mucous cells on the filaments, there are beside the rectum two very large mucous glands which consolidate particles before they leave the cavity. Perhaps inappropriately, from their position, these are called *hypobranchial glands*.

The mouth is carried on the snout close to the ground. The early mollusc feeds by rasping up small particles and raking them into its buccal cavity by an organ called the *odontophore*. This is worked by a complex of small muscles and is a unique feature of the molluscs. It forms a broad tongue covered with a chitinous membrane, the *radula*, bearing very numerous teeth. The odontophore is alternately spread out and licked over the substratum and then withdrawn, taking its load of food into the mouth like a scoop. Mucus is provided by small 'salivary' glands and by the wall of the oesophagus. A series of small food boluses passes down the gut, bound into a continuous mucous rope as they go. The stomach of the early mollusc[155] is rather complicated. The food rope, as it passes in, becomes continuous with a stiff mucous rod which projects into the stomach from the intestine and is continuously rotated by cilia. This rod has been called the 'protostyle' and forms a windlass which helps to draw in the food from the oesophagus. As the string is rotated on the protostyle it is swept repeatedly over the ciliated wall of the stomach. Particles detached from it are here graded for size by the ridges of a *ciliary sorting area*. Small nutritive particles at length pass into one of the paired openings of the *digestive diverticula*, leading to the tubules of the bulky digestive gland. Here they are phagocytosed and digested within the cells. Coarser particles are plastered on to the protostyle or by-pass it into the intestine. Pieces of the rod itself are nipped off by muscular contractions at the far end and moulded into faeces as they pass back to the rectum.

The early mollusc is shown in Fig. 1 as having a heart consisting of a median ventricle and two lateral auricles. Into each auricle blood passes directly from the gill of the same side, and from the ventricle it is sent to the whole body by closed arteries. The true *coelom* is a very small space, just large enough to provide a pericardium round the heart, and to form a gonadial cavity in front of the pericardium. These two spaces communicate, and the eggs and sperms shed from the wall of the gonad pass into the pericardium. They are removed by paired *coelomoducts* which open into the mantle cavity, and these ducts have glandular walls which extract nitrogenous waste from the blood. In addition, the pericardial wall is itself excretory. Protonephridia with flame cells are found in molluscs only in some embryos; the permanent excretory organs are always open coelomoducts.

Most of the space round the viscera is not coelomic, but is filled with venous blood which drains there after transport round the body. This blood-filled space reaches into the spongy muscular meshwork of the head and foot, and may be referred to as a *haemo-*

coele, though this is a rather imprecise term. There is no evidence that the molluscan haemocoele was formed, as it is in Arthropoda, by true blood vessels encroaching on a large coelom. At all events, where the annelids and lower worms make use of coelomic fluid as a hydroskeleton, the molluscs employ the great volume of blood in the haemal spaces. This forms a malleable 'haemoskeleton' that can be manipulated by the muscles of the body wall. By appropriate shifting of blood the mollusc performs startling changes of shape. The foot is dilated, the proboscis extended or the whole head region enlarged as the animal expands from the shell.

In an early mollusc such as a chiton, the nervous system is extremely simple, reminiscent of the flatworm level. The central portion is a ring surrounding the oesophagus, formed by a dorsal *cerebral* and a ventral *labial commissure*. Two pairs of parallel cords run back from this ring, *pleural cords* lying laterally and innervating the organs in the mantle cavity, and *pedal cords* running along the foot. Cross connections link these cords in ladder fashion. The cords themselves are studded with nerve cells, but the only neurones organized at this stage into ganglia are those of the buccal ganglia which control the odontophore. For the rest, apart from the muscles that withdraw it into the shell, the early mollusc is a slow-working creature, with little fast nervous control. Mucus and cilia serve most of its needs. The muscular parts are the buccal mass and the foot, but the general dominance of nerve and muscle is yet to come.

Sense organs are distributed at the points most exposed to stimuli. The paired tentacles on the head are tactile and probably gustatory. Eyes are formed by simple pigmented retinal cups. Small otocysts are embedded in the foot at the base of the nerve rings. An osphradium lies – as we have seen – at the base of each gill, forming a pallial sense testing the entering water current. Around the base of the foot runs an *epipodium*, a fringe like a false mantle, bearing delicate tentacles that come in contact with the substratum. They are probably both chemosensitive and tactile.

Fertilization is external. The eggs have no protective capsules and little yolk. They hatch after spiral cleavage into top-shaped *trochophore* larvae, with a ring of cilia, very like the larvae of annelid worms. A further larval organ soon appears, a wheel-shaped *velum*, fringed with long cilia. A foot develops upon the ventral side and a shell-gland near the lower pole soon secretes a horny larval shell.

In all its primitive features, our 'early mollusc' obviously draws close to the class Monoplacophora. Our insights into early structure were greatly illuminated by the finding of living *Neopilina* by the Danish Galathea Expedition. In the most momentous discovery in the history of malacology, Dr Henning Lemche announced in 1957 a mollusc previously known only as limpet-like fossils from the Cambrian to the Devonian. Dredged from 5000 metres off the Pacific coast of Mexico and named *Neopilina galatheae*, this new mollusc has a flat, saucer-shaped shell up to 4 cm long, and a ventral foot, a mouth in front and the anus behind (Fig. 2). The gills, lying in shallow grooves between the mantle skirt and the foot, were found to number five pairs! Still more remarkable, the auricles, renal organs and gonads are multiplied too. The muscles attaching the animal to the shell number eight on each side (some fossil monoplacophoran shells show different totals of muscle scars). Though a second living monoplacophoran, *Vema*, was reported in 1959 from off Peru, the important work is still the monograph by Lemche and Wingstrand on the morphology of the fixed Galathea material.[189]

Appearing in 1960, this account was eagerly awaited, both for the first full description of *Neopilina* and for the details that would shed light on molluscan origins. The head carries two sorts of appendage, a transverse pre-oral flap, homologized with the velum of the molluscan larva, and a pair of post-oral tentacular tufts, comparable perhaps with the labial palps of bivalves or early cephalopod tentacles. Each of the five gills on either side is capable of muscular movement, and bears on one side a row of finger-like lamellae. Each row of the radula bears a median tooth and five laterals at either side, of which the fourth is comb-like, strangely resembling that of some chitons. The stomach has a simple style sac and contains small particles of ooze especially rich in radiolarians.

There are six pairs of renal organs. The coelom consists of a pericardium around the heart, and paired dorsal coelomic sacs. These sacs correspond in position with the gonad of other molluscs, but in *Neopilina* they are sterile and the gonocoele lies ventrally. The renal organs are branched into lobules and have narrow connections with the dorsal coelomic cavities. As well, the third and fourth renal organs receive in the female ducts from two interdigitate pairs of ovaries and in the male from two pairs of testes. These renal organs may be filled at times with gametes. The heart has a median ventricle and the last gill is drained by a second auricle at either side, the first auricle serving all the others. The nervous system resembles that of chitons, having no ganglia except at the head end, and the pedal and pleurovisceral cords are united by ten pairs of cross connections.

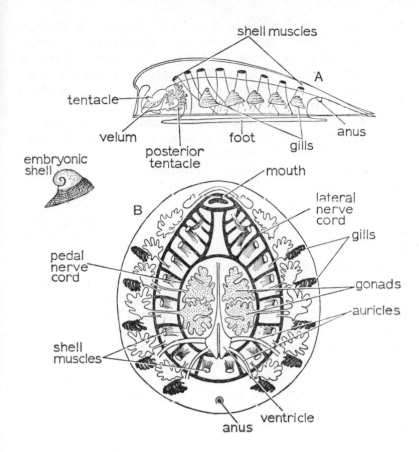

Figure 2 Monoplacophora
(A) Lateral view (*after Lemche*)
(B) Schematic view of internal structure (*after Lemche and Wingstrand*)

Origins and radiation

Over the past twenty years a wealth of new facts have come to bear on our knowledge of molluscan origins and history: not only from *Neopilina*, but from the hitherto unrealized significance of the worm-like molluscs – even more archaic than Monoplacophora – beginning to the class Aplacophora.

General consensus would now go back to the turbellarian flat-worms to reconstruct the features common at the outset to most of the higher metazoan phyla. These must include incipient segmentation, a complete digestive tract, and a trochophore larval stage. Such an early 'prostomatous' stock must have given off the existing Turbellaria as a side-branch. The origin of the molluscs may not have been far from these: probably near the 'pre-annelids' in small ciliated acoelomate worms, moving by ventral ciliation, or with a longitudinal wave over the ventral surface.

The molluscs – it would appear – never achieved segmentation, at least not with the arithmetical constancy shown by the thorough metamerism of annelids and arthropods. Nor were molluscs ever in a proper sense coelomate, though – as they increased in size – they must early have abandoned locomotor support by a solid body wall, and developed the labile form and fluid skeleton offered by blood in an extensive haemocoele. Such early metamerism as began to appear in several groups has declined, to be wholly obscured or lost in each of the higher classes.

At the outset the molluscs must have acquired their distinctive growth gradients. The general contractile rhythm of the turbellarians would then have been confined to the ventral muscular expanse, which became alone concerned with locomotion. The dorsal part of the body, lying above and carried upon the locomotor surface, became progressively visceralized and covered by the mantle and shell. With the loss of participation in locomotion the 'visceral mollusc' – if we may so call its dorsally expanding part – lost its antero-posterior orientation, and began to obey fundamentally different laws of growth and symmetry from those of the 'somatic mollusc' (the head and foot) that carries on the primitive locomotor function inherited from the flatworm level of organization.

In one of the best modern reviews of early molluscan organization, Stasek[282] postulates the early stages leading to a recognizable modern mollusc, from a worm-like acoelomate creature, only a few millimetres long, moving by cilia upon a mucous track. Among its attributes were a diverticulated gut, a sub-terminal dorsal anal pore, longitudinal nerve cords, a series of dorsoventral muscles, and a pair of mesodermally derived excretory tubes, dorsolateral to the gut (see Fig. 4).

At an early stage – it is next assumed – a dorsal cuticle developed, with microscopic spicules laid down in it, as in the skin of Aplacophora and the chitonid girdle today. A later deposition of calcium carbonate beneath would have raised this general cuticle to form an overlying periostracum, still thin, and now secreted by a marginal periostracal groove. The theory of Wilbur for calcareous deposition in the molluscan shell still requires the preliminary secretion of its matric in horny conchiolin. Raven has shown this also happens in ontogeny.

The projection, with overlap, of the shell and its underlying skin-fold, the mantle, sets up a mantle cavity and allows the foot an increased independence and mobility beneath. With larger size, the distribution of the blood, now both a respiratory and a 'skeletal' medium, would require the elaboration of a haemocoele, and pumping by a heart. Gills would appear first as simple vascularized out-pushings of the body wall into the pallial space.

The molluscs acquired a truly coelomic space only incidentally, by the appearance of a pericardium around the heart with the expansion and fusion of mesodermal (probably originally gonoducal) tubules. Without this, they would be regarded today as acoelomates, having, says Stasek, 'escaped that designation on a technicality'.

Associated with the early pericardial space were also gonoducts, with the original gonadial cavity widely continuous with the peri-cardial coelom itself. The mesodermal tubules thus included one or more pairs of renal ducts, and direct, separately opening, gonoducts.

While these transformations were happening, the buccal mass must have been taking on its sweeping or abrading role in food collecting, with the emergence of that widely unifying molluscan organ, the odontophore with radula. Another feature, inevitably giving a pseudo-segmental appearance, was the series of paired, dorsoventral muscular bands, now becoming inserted on shell scars and complexly radiating into the foot, where they distribute upon the sole.

Derivation of the classes

The important researches of Salvini-Plawen[278] have recently taught us much about the ancestral significance of the important shell-less stem-group, the worm-like Aplacophora. This class has no littoral members; they live in moderate to abyssal depths. The spiculose dorsal mantle is enrolled to envelop the narrow worm-like body, completely so in the sub-class Caudofoveata, but only partly in the sub-class Ventroplicida leaving a narrowly restricted ventral pedal groove.

Both sub-classes possess the molluscan hallmarks of a simple radula within the buccal cavity and an elementary posterior mantle cavity. They have also numerous serially arranged pairs of dorsoventral muscles. Their 'test-cell-larva', to be described later, seems to be a primitive feature shared with chitons, scaphopods and protobranch bivalves.

The old name 'Amphineura', for the Polyplacophora (chitons) and the Aplacophora together, should now be dropped, for Salvini-Plawen has shown that Aplacophora do not actually possess two separate pairs of medullary nerve cords, as do chitons and also *Neopilina*. But Salvini-Plawen's two groupings Ventroplicida (Solenogastres) and Caudofoveata need hardly be raised to separate class level. With Stasek[283] we prefer to retain the single class Aplacophora that can be placed in a sub-phylum Aculifera, along with Polyplacophora, forming a second class, leaving the Monoplacophora and its 'four descendant sub classes' as a second sub-phylum Conchifera. Within the Aculifera, the Aplacophora and the chitons probably stand rather far apart.

The Ventroplicida (*Neomenia, Proneomenia, Paramenia*, and *Lepidomenia*) live suctorially, twining upon the bodies of gorgonians and hydroids. They respire through the general epithelium or by simple skin-folds of the pallial cavity. The Caudofoveata burrow peristaltically in organic muds and ooze. The foot is reduced to a small post-oral lobe, and the bell-shaped posterior mantle cavity has a pair of ciliated gills.

The Polyplacophora or chitons are pseudometameric, flattened like limpets, and highly specialized for adhesion to a hard substratum. The head and mouth are anterior, the anus posterior. The shell is represented by eight transverse dorsally arched valves. Round the margin and fitting closely to the ground runs a fleshy or horny girdle, provided with spicules or scales. The close-fitting habit of chitons (rolling up like a wood-louse when detached) has led to great modification of the mantle cavity, which now reaches forward to the head as narrow side-grooves lying between the girdle and the foot. The gills are clearly primitive ctenidia, but are secondarily multiplied into long series, extending progressively further forward until in the higher chitons they reach the head. An inward water current can be admitted at any point by temporarily raising the girdle. After bathing the gills, the pallial current passes out in the posterior mid-line, along with faeces and excretory products.

The Polyplacophora are a compact class, of some six families, uniformly herbivorous (with the strange exception of *Placiphorella*, that has acquired a carnivorous habit). The digestive, reproductive, excretory and nervous systems remain highly primitive.

Beedham and Trueman[58] have studied the cuticle of *Proneomenia* and compared it with the spiculose girdle of the chiton *Acanthochitona*. The aplacophoran cuticle is equated with an early mucoid stage of the molluscan shell. The secretion of additional protein, with quinone-tanning, would seem necessary before a calcified shell evolves. The spicules are similar, but differently distributed in the two classes. The girdle in chitons would seem to represent the aspiculose mantle fold in Aplacophora, whose spiculose cuticle in turn becomes homologous with the aspiculose periostracum of chitons.

The Monoplacophora

Unlike the chitons, all the pre-Cambrian Monoplacophora have a single shell-piece, and they would seem to be the parental stock from which all gastropod (and probably other) lineages have sprung. By the lower and middle Palaeozoic they had become very diverse, but have today only six or seven surviving species, all belonging to the one genus *Neopilina*. In the lower Cambrian, the chief monoplacophorans were the Palaeacmaeidae (12–15 mm long), cap-shaped and with six pairs of muscle scars. Stasek has speculatively brought them closer to the molluscan stem by endowing them with true ctenidia, unlike the modified gills of *Neopilina*.

In late Cambrian, two sub-classes of Monoplacophora diverged. The Cyclomya emphasized the dorsoventral growth axis, and the muscle scars are reduced, with their insertions well within the aperture, by inference allowing the animal to withdraw fully into the shell. The heightened mantle cavity – it is suggested – reduced the ctenidial number to one pair, and the head could now wander freely about and project from beneath. Extinct by the Devonian, the Cyclomya could earlier – with visceral torsion – have given rise to a gastropod derivative.

The second sub-class, Tergomya, were elongate and relatively flat. They gave rise to no innovative offshoots, but include the modern *Neopilina*, securely surviving perhaps by virtue of its ecological seclusion as a specialized ooze feeder in deep water.

The gastropods could have been derived from cyclomyan monoplacophorans by the head evolving a freedom as the mantle cavity moved forward to overarch the neck, and the pedal muscle insertions withdrew deeper into the shell. Thus, the 'head-foot' became separated by a narrow waist from the visceral mollusc; and it is this

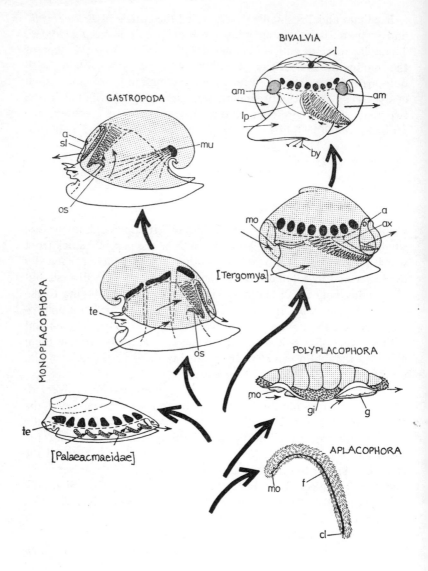

Figure 3 Derivation of Gastropods and Bivalves in relation to early molluscs (*based on Stasek*)

a, anus; am, adductor muscle; ax, gill axis; by, byssus; cl, cloaca; f, foot; g, gill; gi, girdle; l, ligament; lp, labial palp; mo, mouth; mu, shell muscle; os, osphradium; sl, shell slit; te, tentacle modification that allows 'torsion', the special developmental hall-mark of the Gastropoda, to take place.

The Bivalvia, so obviously divergent from the gastropods, could have been produced from Monoplacophora in which the mantle cavity was already raised, with the heightening of the visceral mass. The molluscs here enter the adaptive zone of filter feeding, already occupied before them by sponges and brachiopods. During the flattening and lateral compression of the body, with its gain in height, two separate lateral centres of calcification appeared, united by an intermediate, nearly uncalcified band of conchiolin.

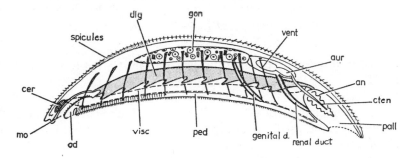

Figure 4 Hypothetical molluscan derivative from unsegmented flatworm an, anus; aur, auricle; cer, cerebral ganglion; cten, ctenidium; dig, digestive gland; gon, gonad; mo, mouth; od, odontophore; pall, pallial cavity; ped, pedal cord; vent, ventricle; visc, visceral loop

The Gastropoda and torsion

As well as being the largest class of molluscs, the Gastropoda are easily the most varied. They crawl primitively by a flattened foot attached to the ventral surface which in many ways keeps close to the original mollusc pattern. Most gastropods, including the earliest, have the shell and the visceral mass coiled in a right-handed spiral which they carry dorsally (Fig. 3). With this spiral coiling, many students have confused the process called *torsion* of the visceral mass.

Torsion is a very important event in gastropod history, and is quite independent of spiral coiling; some gastropods are not coiled at all, but all gastropods – at some time in phylogeny – have undergone torsion. We must try to explain how torsion occurs and what it means in the life of the mollusc. It is a change which brings the mantle cavity to the front of the body, while the visceral and pallial organs are

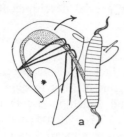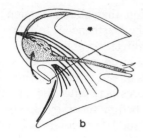

Figure 5 Veliger of *Patella* (a) before and (b) after torsion, showing the relation of the asymmetrical shell muscle to the gut and the movement in position of the mantle cavity, during the passage of the visceral mass through 180° (*modified from Crofts*) *mantle cavity

altered in position by twisting through 180° in relation to the head and foot. From the diagram in Fig. 5 it can be seen that the mantle cavity before torsion faces backwards and ventrally. The larva has at this stage an asymmetric retractor muscle attached to the shell on the right, some of whose fibres sweep dorsally over the gut to be inserted in the left side of the head and foot. When this contracts at the beginning of torsion, the bulky visceral mass is pulled over from above towards the left. It is slung round the left side of the animal to a ventral position, and the mantle cavity at the same time moves up the right until it is dorsal and faces forwards.[225] Everything behind the 'neck' – which is the actual site of twisting – is thus reversed in position. Thus the gill, auricle and kidney that were at first left are now right, and vice versa. At the point of twisting, the two long nerve connectives running to the viscera are crossed in a figure of eight.

Torsion is a drastic process that would be impossible during adult growth. Yet in the larvae of *Acmaea* and *Trochus* its first stages through 90° take only a few hours, as the visceral mass is slung over to its new position by contraction of the asymmetric muscle. Later stages of torsion are added by asymmetric growth of the larva.

What is the value of torsion? More controversy has been devoted to the adaptive explanation of this change than to any other point of molluscan biology. The theory most favoured today and involving fewest improbabilities is Garstang's. He held that torsion first took place as a larval mutation, of little direct use in the adult. Before torsion, when the veliger retreated from predators into its shell, the posterior mantle cavity could receive the head and velum only after the foot was already inside. With the mantle cavity to the front, the sensitive parts were no longer exposed; the head and velum could withdraw first, followed by the foot and later an operculum to seal the aperture. Such an explanation of torsion serves well enough if we are thinking of predators of medium size like chaetognaths or stinging coelenterates. But what of larger fish, such as herrings and mackerel, which must be the heaviest eaters of larvae and may swallow hundreds of veligers at one draught?

It is doubtful, too, whether the needs of the adult should be ruled out so completely. The initial value of torsion to the larva is obvious. But the pallial cavity is such a commanding feature of the adult that it cannot be an indifferent matter whether it opens in front or behind. In a freely moving gastropod it would seem moreover highly advantageous that the mantle cavity should be at the advancing end. This would convert it from primarily a cloaca with gills into an organ in sensitive touch with the environment of the head. As the animal orients by the sense organs of the head and makes continual small adjustments of position, the pallial opening would be so placed as to make immediate contact with every slight change of surroundings. The gills could now be bathed by undisturbed water from in front of the animal, and the osphradium could continuously sample the environment into which the animal was moving.

After torsion, the anus lay in front and the animal could not conveniently grow longer. The viscera, including the loop of the gut, began to bulge in a dorsal hump, which could in turn be disposed most compactly by dorsal coiling. In most living gastropods, the viscera have thus formed a helicoid, right-handed spiral. As the penultimate whorl of the spiral bulges into the right side of the mantle cavity, a new asymmetry is produced there and one set of pallial organs tends to be reduced and lost. To this we shall return later.

The Scaphopoda

The tusk shells are burrowing molluscs with a tapered shell, as if the conical cap of the early molluscs had been produced upwards into a slender tube open at either end (Fig. 6). The shell projects obliquely from the sand, the broad end lying deepest and containing the head and foot. The foot is plug-shaped and can be extended and plunged into the sand, drawing the animal down by subsequent contraction. The head bears several bunches of slender ciliated tentacles called *captacula*. These reach into the substratum to bring back to the

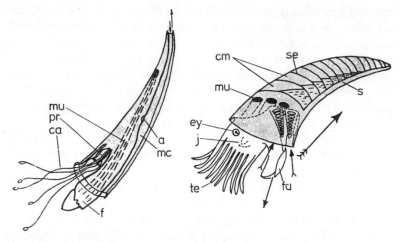

Figure 6 Scaphopod (*left*) and hypothetic early Cephalopod
a, anus; ca, captacula; cm, camerae; ey, eye; f, foot; fu, funnel; mc, mantle cavity; j, jaw; mu, shell muscle; pr, proboscis; s, siphuncle; se, septum; te, tentacles

mouth small particles such as foraminifera.[226] The buccal mass has a large strong radula, and the stomach is simple, with paired digestive diverticula.

The mantle forms a complete tube, and the pallial cavity, on the ventral side of the body, runs right through the shell. Both inward and outward ciliary currents pass through the posterior end. There is no gill, respiration taking place through transverse folds in the lining of the mantle. The blood circulates in a system of rudimentary sinuses, with merely a contractile portion near the anus serving as a heart. A pericardium is lacking, but there is a pair of small renal organs opening at either side of the anus. The sexes are separate, and there is a large median gonad, discharging its gametes through the right renal organ.

The Bivalvia

The lamellibranchs or bivalves are in some ways the most highly modified of all the molluscs. They have completely lost the head, the buccal mass and the radula. They are a rather more uniform class than the Gastropoda, and the great majority are ciliary feeders with extreme development of the gills. The mantle cavity is much larger than in other molluscs and has quite dominated lamellibranch evolution. Two symmetrical mantle flaps enclose the whole body, and these secrete right and left shell valves hinged in the dorsal line. The shell can be tightly closed by the *adductor muscles*, which are the shell retractors of the early mollusc rearranged to run directly between the valves.

With few exceptions all lamellibranchs are sedentary. Some of them remain anchored to the substratum, or firmly cemented, as with oysters. Those that burrow do so by the use of the foot, which is usually a compressed muscular tongue that can be elongated and thrust forward in the substratum to haul the animal behind it.

The lamellibranch gills remain paired, and each consists of two plate-like *demibranchs* hanging in the mantle cavity at either side of the foot. In most bivalves their filaments have become bent back upon themselves like a V, so that each of the demibranchs is double and composed of two flat *lamellae*. The cilia of the gill draw a powerful water current into the mantle cavity; a typical lamellibranch filters thirty to sixty times its own volume of water in an hour. As the water passes between the filaments, food particles are held back by straining cilia and are carried forward by other cilia in mucous strings towards the mouth. Here they are carefully graded for size by ciliated labial palps before being ingested. Cilia and mucus are just as important in the gut as in the mantle cavity. The stomach is extremely elaborate, with large ciliary sorting areas and a long rotating style. Specialized as they are in the pallial and digestive organs, the bivalves are much less so in the reproductive, circulatory and excretory systems. These are but little altered in plan from those of early molluscs.

The pallial and visceral organs having developed at the expense of the head, most of the sense organs have quite withdrawn from the anterior end. The margin of the mantle is now the site of contact with the environment, and develops abundant tactile organs and in some cases, as in the scallops, eyes as well.

The Cephalopoda

The earliest undoubted cephalopod is the small, slightly curved *Plectronoceras* of late Cambrian (see Fig. 6), with a few wide-spaced septa and a large siphuncle along the concave side. Yochelson has suggested for it a derivation from moderately high, conical Monoplacophora. *Kirengella* (late Cambrian) significantly has the widest part of its oval section towards the front, and four pairs of muscle scars. *Knightoconus* lacks the scars, but is moderately high, of oval cross section like *Kirengella*, and curved backwards. It has moreover several septa that form chambers in the smallest part of the shell, and – if endowed with a siphuncle – could have bridged the way to *Plectronoceras*.

The Cephalopoda are not only the most elaborately evolved molluscs, but are entitled to pride of place among all the invertebrates. They are almost all fast-moving carnivores, and either pelagic or at least much more independent of the bottom than other molluscs. The class is named from the close union of the head with the foot, which has become much subdivided to produce two new types of organ. First there is a series of prehensile tentacles which have spread completely around the head so that the mouth now lies at their centre. In primitive *Nautilus* (Fig. 21) these tentacles are small and very numerous. In higher cephalopods (Fig. 22) they form a set of eight or ten muscular arms, bearing long rows of suckers. The octopods have eight long arms equally developed, and the cuttlefish and squids a circlet of eight short arms and two long tentacular arms.

The second organ developed from part of the foot is a muscular spout or *funnel* lying behind the head at the posterior side of the animal, where the mantle cavity is situated. This controls the exit of water from the mantle cavity and produces a strong jet which is used in swimming. In its proper orientation, the body of a modern cephalopod is elongated dorsoventrally, the ventral surface carrying the mouth surrounded by the tentacles, and the opposite end – blunt or pointed – being dorsal. In normal swimming, the posterior surface, with the funnel, is the lowermost, and the anterior surface is carried above.

The shell plays only a subsidiary role in modern cephalopods, being internal or even lost. Most extinct cephalopods, however – as well as the ancient living *Nautilus* – have a large external shell. This is closely coiled in a plane spiral with the hump carried dorsally, and is divided by septa into successive chambers filled with gas. They are quite closed off from the animal, which occupies only the last one, and by their lightness they confer buoyancy.

In the modern cephalopods – where the shell is internal – the visceral hump is covered by a muscular mantle, giving the body a rounded or a streamlined contour. In squids the aboral end is tipped by a pair of fins, and in cuttlefish these run along the sides of the body. The ciliary functions of the early mollusc are now assumed by muscles. The mantle cavity contains two gills, but these are no longer ciliated and water is pumped in and out by strong contractions of the muscular mantle. With the loss of the external shell the whole mantle can freely contract, and jet propulsion is the characteristic means of locomotion. In deeper-water forms the swimming function may be taken over by the pulsations of a web running between the arms.

The most prominent sense organs of cephalopods are the eyes, which reach a perfection found nowhere else in the invertebrates. The brain is very elaborate too. The nerve centres of the primitive mollusc are concentrated within a cartilage head capsule, and the cephalopods have acquired many of the higher functions of the brain as found in vertebrates. Speed, alertness and large size are the keynotes of the Cephalopoda. Like the bivalves they have concentrated with great success upon one possible molluscan pattern. Yet such are the differences between, for example, a clam and a squid that it is hard to realize that the resources of one phylum have made both of them possible.

The shell

The molluscan shell is formed by the deposition of crystals of calcium carbonate in an organic matrix of a protein substance called con-

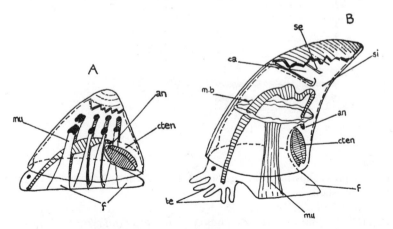

Figure 7 Reconstruction of monoplacophoran *Kirengella* (A) and Cambrian cephalopod *Plectronoceras* (B)
an, anus; ca, camera; cten, ctenidia in mantle cavity; f, foot; m.b, muscle band; mu, shell muscle; se, septum; si, siphuncle; te, tentacles

chiolin. Its structure is highly ordered and falls ideally into three zones: (1) an outer organic layer or periostracum, thought to be a quinone-tanned protein, (2) a prismatic or columnar crystalline layer, of calcite in some species, aragonite in others, added to along the edge of the mantle, and (3) an inner layer, sometimes nacreous, formed of thin, laminate crystalline sheets and produced by the mantle's inner surface (Fig. 8).

Calcium supply is taken up with the food from the sea-water environment, and land and freshwater shells, where calcium may be deficient, are often very thin. Carbonate is derived from $CO^2/$ bicarbonate pool available in the mantle or general body tissues. Shell formation takes place in the thin layer of extrapallial fluid lying between the mantle and the inner shell surface. Here the two solid phases of the shell – organic and crystalline – are formed; and the composition of the fluid determines the chemical nature and pattern of the matrix, and the rate and manner of crystal growth. The organic matrix is protein in nature, and is shown by the electron microscope to be a fenestrated sheet with a characteristic pattern of holes at which crystal growth takes place.

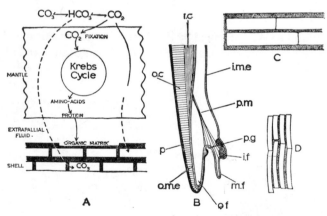

Figure 8
(A) Diagram showing relations of CO^2 to calcium carbonate and organic matrix of shell
(B) Section of shell and mantle edge
(C), (D) Sections of prismatic and nacreous layers (*all after Wilbur*)
i.c, inner crystalline (nacreous) layer; i.f, inner mantle fold; i.m.e, inner mantle epithelium; m.f, middle mantle fold; o.c, outer crystalline (prismatic) layer; o.f, outer mantle fold; o.m.e, outer mantle epithelium; p, periostracum; p.g, periostracal groove; p.m, pallial muscle

2 External form and habits — Gastropoda

A typical snail owes its outward shape to the shell and the flattened foot. The shell forms a cone coiled in a helical spiral, and its lowest and widest coil is the spacious body whorl, into which the animal can retreat. The foot is drawn in last, and in the first sub-class of gastropods, the Prosobranchia, the aperture is usually sealed by an operculum, a horny or calcareous plate borne on the back of the foot. In the limpets, however, the animal never emerges from the shell, there is no operculum and the foot is permanently attached to the ground. The shell is here a flattened cone, drawn down upon the animal by a circle of muscles arising from the foot. In more active prosobranchs the head and foot are attached to the viscera by a narrow 'waist'. They can be quickly withdrawn into the shell by the columellar muscle, springing from the foot and attached to the axial pillar of the shell.

The foot is held to the ground by a viscid mucus. Its sole is ciliated and some smaller snails glide along simply by cilia working through the mucous layer as in some of the flatworms. Tiny marine snails such as the Rissoidae may haul or suspend themselves through the water with a mucous thread secreted from the pedal gland.

Most gastropods, however, creep by a succession of continuous waves of muscular contraction passing over the sole from behind. The most detailed analysis of gastropod locomotion has been made by Lissmann,[194] using for experiments *Helix*, *Pomatias* and *Haliotis* (Fig. 9). In *Helix* we find a good example of a direct locomotor wave, where a pattern of contractions followed by relaxations passes over the longitudinal muscles of the sole from behind forward. The whole sole thus shows at any one time eight or ten dark bands moving forward. These indicate phases of foot movement, where longitudinally contracted cross-furrows are lifted off the ground and shifted forward. Between the bands are areas of relaxed longitudinal muscle that remain at rest.

In the land operculate snail *Pomatias elegans* there are no succes-

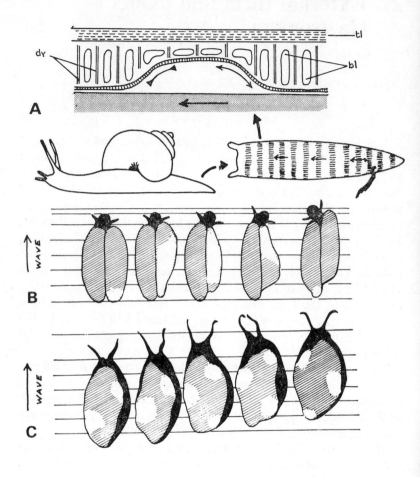

Figure 9 Gastropod locomotor waves
(A) *Helix* with locomotor wave pattern of sole, and (*above*), musculature of a single raised area of the foot
(B) *Pomatias* stepping with one half of the foot at a time
(C) Movement of *Haliotis*, with raised parts of foot shaded
bl, blood spaces; dv, dorso-ventral muscles; tl, transverse and longitudinal muscles

sive waves, but a virtually bipedal locomotion, the right and left halves of the foot alternating in state and being demarcated down the middle line. One half is raised, then contracted behind, next extended forwards and then put down again. The other half is then lifted, posterior end first, and goes through the same contraction behind, extension forwards and placement in advance. This movement has been compared to that of two feet shuffling forward in a sac.

The ormers *Haliotis* show also an alternate locomotor pattern, but with a wave length intermediate between *Pomatias* and *Helix*. The foot is active and agile, and at any one time has three areas in each half, two attached and one moving forward, balanced with one attached and two in motion on the other side. Just as with six legged insects, a triangular cross-balance of three areas of movement and three of attachment gives both bilateral stability and a ready facility of turning.

Lissmann has analysed in some classic papers the mechanisms – kinetics and kinematics – of gastropod locomotion. In *Pomatias*, the simplest case, there are two antagonistic forces, one elongating and the other contracting. When a wave of contraction begins from the back of the foot, this part is lifted off the ground and drawn forward on the anterior region, still affixed to the ground. As contraction passes forward, the adhering area is reduced, and the whole half foot eventually lifted off the ground. Relaxation of the lifted half now begins posteriorly, and the posterior edge of the foot is lowered. As soon as sufficient surface is in contact with the ground again, adhesion is secured. Internal pressure, or other force of elongation, is increased, while the force of contraction decreases. The foot has regained its original length but its position in relation to the ground is now advanced. The posteriorly directed component of the force of extension is now resisted by the adhesion of the back of the foot. The anterior part is pressed forward with extending force sufficient to overcome any frictional resistance.

In *Helix*, each of the numerous transverse furrows is undergoing extension in front and active contraction behind. Each single functional unit is in essentials the same as the whole half-foot of *Pomatias*. The external forces of dynamic friction and static reaction were measured by Lissmann, as the snail crawled across a movable bridge, mounted on a knife edge, to record tensions or thrusts between the fixed platform and the bridge. Static reaction/sliding friction curves are shown in Fig. 10, both for a single fixed area and for the alternation of separate stationary and moving areas.

B

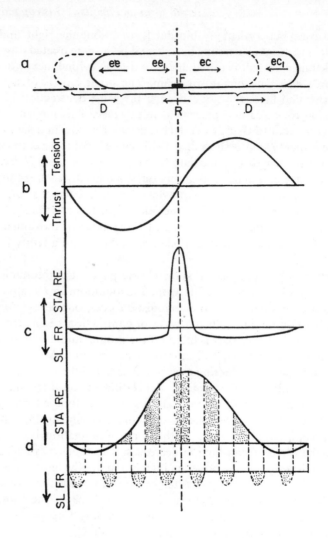

Figure 10   Kinetic effects in snail locomotion

(a) Mechanical effects of foot compared to a rubber balloon, with its central region attached, posterior end contracting and anterior end relaxing. Movement as shown by the dotted outline, giving a thrust/tension curve as in (b)

(c) Static reaction/Sliding friction curve for a single fixed area, and (d) from the alternation of separate stationary and moving areas

D, dynamic friction; F, area of fixation; R, static reaction; ec, ec_1, external force of contraction; ee, ee_1, external force of extension

Prosobranchia

The prosobranchs are immensely varied in structure, and the architecture of their shells usually has an obvious adaptive meaning. Shell shape is almost always a poor guide to natural relationships and there are certain types of shell we shall find repeatedly in each of the three orders which make up this sub-class. These orders are the Archaeogastropoda, the oldest and least specialized; the Mesogastropoda, the largest order of molluscs, and particularly diverse in mode of life; and the Neogastropoda, which are rather specialized carnivores.

Surveying the two higher and larger sections of the prosobranchs, we shall begin with those of the order Mesogastropoda. In their evolution, these are far from conservative.

Wherever molluscs can exist at all they have some representatives, and every adaptation we shall later find in the opisthobranchs and pulmonates, the mesogastropods seem somewhere to have attempted for themselves. They may give rise in several groups to limpets; the best known are the Calyptraeidae or Chinaman's hat, cup and saucer and slipper limpets, slow-moving or entirely immobile and collecting their food by the ciliary currents of the gill. The shell periphery is here drawn out and reaches to the ground, making an originally top-shaped shell into a depressed cone, with distinct upper and lower surfaces. *Calyptraea* still has a small distinct spire, but in *Crepidula* this has vanished entirely, leaving the shell an oval plate with a pocket for the viscera. The Capulidae are a family of small, high-peaked limpets usually attached to bivalve shells and tending towards ectoparasitism. Such habits are taken further in the limpet-like genus, *Thyca*, classed near the Capulidae, an ectoparasite of tropical starfish, and from near this point a whole chapter of mesogastropod parasitism, external and finally internal and degenerated, has been developed.

In some ciliary and mucus-feeding mesogastropods the shell shape has become very aberrant and loosened like a corkscrew or irregularly coiled as with a serpulid worm. These include two separate families. The first, the Siliquariidae, are derived from the auger-like Turritellidae, also ciliary feeders; and in *Vermicularia* the ontogeny betrays phylogeny, the first whorls of the shell forming a normal looking turritellid. In the Vermetidae most species have abandoned ciliary feeding and collect plankton by secreting long mucous strings. They are brightly coloured and *Serpulorbis* lacks the operculum, retreating for protection far up the shell tube.

To return to the main stream of the Mesogastropoda, the fusiform or spindle-shaped spire is as characteristic here as is the top-shaped or limpet-type shell among the archaeogastropods. The shell aperture is prolonged into a spout-like anterior canal, projecting in the line of advance, and the animals are for the most part rather mobile and often carnivorous. The canal is traversed by a pallial siphon bringing water into the mantle cavity, enabling the osphradium to sample the environment ahead, and to make an olfactory search for food.

The earliest and most primitive mesogastropods are perhaps the periwinkles or Littorinacea. They lack a shell canal, and are small and often drab, though beautifully adapted to high-tidal life. In most parts of the world they characterize the supralittoral fringe, but in the Atlantic they reach to the midlittoral. The Cerithiidae are distinguished by their long-spired, trailing shells. Like the littorines they are often high-tidal, though chiefly in tropical mangrove swamps and mudflats (*Terebralia, Cerithium, Pyrazus*). *Rhinoclavis* lives in clean sand. All cerithiids graze upon organic deposits or algal films.

Of the higher Mesogastropoda there are three superb series, with an especially prodigal wealth of tropical species. First, the large carnivores of sand-flats include the Cassididae, the Doliidae and – much modified for burrowing – the Naticidae. Largest and heaviest are the cassids, or helmets and bonnets, mainly in shallow water and engulfing burrowing sea urchins with the long extensible proboscis. The thinner and smoother tuns have an even broader foot than the cassids, with which they enwrap bivalve prey, boring a hole in the shell with an acid gland at the tip of the proboscis. The same shell-boring habit is used by the smooth and polished Naticidae or necklace shells, in which the soft parts spread about the shell. The animal ploughs along or burrows in fine sand, and the shell is partly submerged in the tissues of the broad foot. The propodium is built up into a fleshy head-shield like a sand-plough which reaches back over the front of the shell. Side-flaps of the foot, known as parapodia, spring up to cover the shell laterally, and the whole animal is converted into a flat mucus-coated wedge for sliding over or into the sand.

The cowries or Cypraeidae are the most beautiful and highly prized mesogastropods of coral shores. Their shells are smooth and ovoid, polished with a high glaze. The significance – if any – of their rich and varied colour patterns is still a mystery, as they are covered from view by an expansible integument, often beset with long papillae and sometimes itself handsomely pigmented. The range of species is legion: the large spotted tiger, the camel, tortoise, rat and deer; the

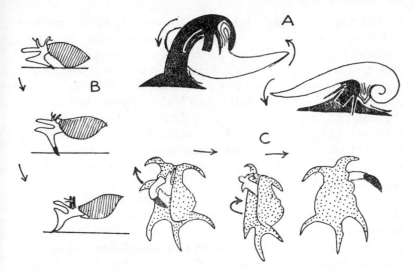

Figure 11
(A) Righting movement of *Aporrhais*
(B) Leaping movement ('backward flip') of *Strombus*
(C) Righting movement of *Lambis* (after Berg)

speckled lynx and thrush, the hump-back *mauritiania*, the chocolate and yellow-banded mole, the snakeshead and dragonshead; the smaller and exquisite *asellus*, *punctata* and *cribraria* (the sieve) and *annulus* (gold ringer).

Not far remote from the cowries are placed the Lamellariidae with soft, slug-like body and thin transparent internal shell: the skin is never conspicuous but camouflaged to resemble the sponge or ascidian background.

The Strombidae or conch shells are a huge and very wide-radiating series, shallowly burrowing in sand or gravel and cropping algae or grazing organic-rich deposits. They are noted for the varying development of the shell lip, a heavy wing in the larger true conchs, a spiny digitate edge in the spider shells, *Lambis*. Others are cone-like or resemble muricids, and *Terebellum* is light and smoothly polished for gliding through the sand.

Frequently, the strombid shell lip has a squint for an optic tentacle, the eyes being spectacular and prominent, at the ends of long peduncles. The animal is active and agile. Strombidae have abandoned the slow creeping locomotion with the sole, and progress by a repertoire of active muscular movements of the foot, probably derived from the 'escape' behaviour developed by prosobranchs of soft flats, in the presence of starfish, cones and other predators. Ordinary locomotion consists of a series of short leaps, throwing the shell forward then subsequently bringing up the foot.[60] 'Righting' is performed by extending the foot stalk over the shell-lip and placing the operculum under the shell, which is then 'flipped' over to bring the dorsal side up. Escape, elicited by the introduction of a predator, is carried out by a 'backward flip' with the shell thrown back by the forward kick of the metapodium, with the operculum thrust into the ground. This accomplished, the animal then moves the propodium rapidly from side to side, and progresses by a series of forward 'runs'.

The most whelk-like of the Mesogastropoda live on hard rocky shores or are to be dredged from deeper water. They include the trumpets and hairy trumpets, Cymatiidae, with fusiform shells, strong axial ribs or varices, often with rich ornamentation and brightly coloured animals. Close to this family are the Murex-like frog shells or Bursidae.

The order Neogastropoda are all carnivores or carrion scavengers with fusiform shells, a strong anterior canal and – as with many mesogastropods – an inhalant siphon forming a roving tentaculate nostril. They may be comprised in four large groups, of which the Muricacea and Buccinacea are typical of rocky ground.

The true whelks belong to the Buccinacea, which include feeders on carrion or fresh meat, but not generally shell-borers. The Buccinidae contain large fusiform temperate whelks, while the related Nassariidae, small, fast-creeping scavengers, have become extraordinarily numerous and varied in tropical sandy muds. The Buccinacea also include the large snails of the family Fasciolariidae such as *Fusus* with its long spiked anterior canal, and the Galeodidae with the giant *Megalatractus* two feet in length and the American whelk *Busycon*.

The Volutacea have no British representatives. They are sand-dwellers, typified by the Volutidae, Mitridae and Olividae. The olives are polished and streamlined for burrowing and the shell is entirely enveloped by the sides of the foot, with a two-piece crescentic head shield and an upright inhalant siphon.

Some interesting but isolated examples of swimming are to be found among benthic prosobranchs, perhaps – like strombacean fast locomotion – as an escape response to starfish and other predators. In the Olividae, *Olivella tehuelchana* and *Olivella verreauxii* swim by wing-like flaps of the metapodium, while *Ancillista cingulata* uses the expanded propodium. The naticid *Polinices josephinus* thrashes

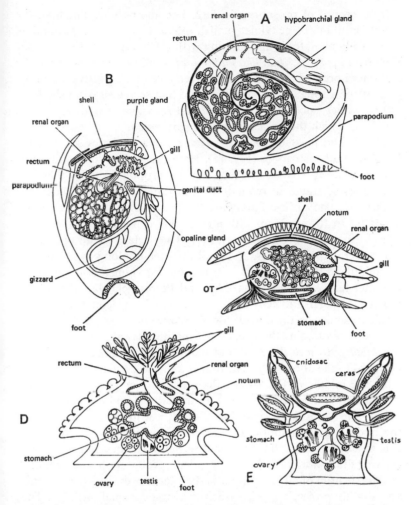

Figure 12 Schematic sections showing the organization of five opistho-branchs

(A) Bullomorph *Haminea*
(B) Aplysiomorph
(C) Pleurobranchoid

(D) Dorid
(E) Aeoliid

the water with its extended propodium. The archaeogastropod *Solariella nektonica* widens the foot to two or three times its normal size, and swims with the shell kept uppermost, but turning it downwards when ready to descend.

Two groups of prosobranchs have become pelagic. The first, the fragile Ianthinidae or 'violet snails', make no attempt at swimming, being truly planktonic and floating with the aperture of the globular shell held upwards.[2] *Ianthina ianthina* may either ride attached to the siphonophore *Velella* – on which it feeds – or may construct a buoyant raft of a tough transparent bubble-like secretion from the foot. Other species, such as *I. prolongata* and *I. exigua*, tow an apron-like raft from which the egg capsules are suspended.

The second pelagic group, the Heteropoda, includes some of the liveliest of all gastropods (Fig. 16). Here the body is light and transparent, and – as we shall find later in the pteropods as well – there is a fine series of stages in reduction of the shell and modification for swimming. The earliest family are the Atlantidae, small transparent snails up to 10 mm long, with a compressed planorboid shell, kept upright in swimming by a sharp keel. They retain a thin operculum and a spiral visceral mass, and scull themselves along by undulating the middle of the sole, which is drawn out into a membranous fin. In the later families, the Carinariidae and Pterotracheidae, the body is much elongated and jelly-like, the viscera being concentrated in a small appendix on the dorsal side. *Carinaria* has a small shell like an elf's cap surrounding the viscera and mantle cavity. *Pterotrachea* and *Firoloida* both lose the shell and mantle, and a naked ctenidium is attached to the dorsal side. The foot is a thin muscular flap, springing from the middle of the lower side, gracefully employed in sculling. In fast swimming the fin side is held uppermost and the viscera hang below. *Carinaria* has a rigid dorsal crest, but in the Pterotracheidae the body is quite smooth, and swimming is aided by lashing from side to side like a small transparent serpent. The buccal mass and its active radula are carried on a pendent proboscis like a trunk. A striking feature of all heteropods is their large eyes, which are tubular and slightly projecting, being freely movable by small muscles. They have a blue metallic sheen, the rest of the body being colourless but for some high spots of purple or magenta.

Opisthobranchia

There are three broadly different types of opisthobranchs: those that burrow in the substrate and possess thin external shells, those that are flattened, naked and slug-like, often beautifully coloured and externally symmetrical, and those that swim (Figs. 15, 16). Shelled opisthobranchs are the earliest and belong mostly to the order

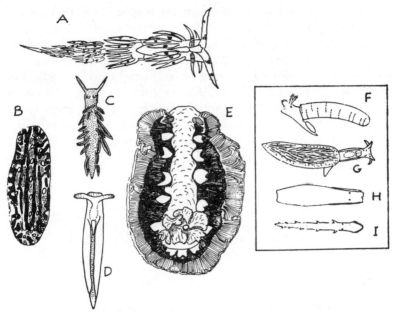

Figure 13 Opisthobranch body forms
(A)   Aeoliid (*Cerberilla*)
(B)   Arminid (*Phyllidia*)
(C, D) Sacoglossans (*Hermaea* and
      Elysia)
(E)   Dorid (*Hexabranchus*)

(*Inset, right*) interstitial sand-
    dwelling gastropods
(F) The prosobranch *Caecum*
The opisthobranchs
(G) *Microhedyle*
(H) *Philinoglossa*
(I)   *Pseudovermis*

Cephalaspidea, alternatively called Bullomorpha from the thin,
bubble-like shell. These must first have evolved along similar lines to
burrowing prosobranchs. They have a wide rectangular foot and a
broad head shield, formed here not by the building up of the front
of the foot, but by expansion of the head itself. Parapodia grow up at
the sides of the body and convert it into a dorso-ventrally flattened
wedge for sliding through or into the sand. Streamlining is carried
much further than in prosobranchs: the shell and visceral hump are
even more reduced and quite incorporated within the new lines of
the body. In early bulloids, such as *Haminea*, *Scaphander* and *Philine*
the head shield is very large, and such excrescences as tentacles, snout
and siphon are drawn close to the outline of the body, with the penis
smoothly invaginated into the head. The head shield tends to push
the mantle cavity further back, and it may even be this that initiated

the reversal of torsion. At all events the mantle cavity moves back along the right side during the history of the opisthobranchs; and before it reaches its original posterior site it has been quite lost, followed soon by the shell and the ctenidium. *Actaeon* and *Ringicula* are the only bulloids with strong or sculptured external shells into which the body can withdraw. *Actaeon* alone has an operculum. A progressive series in slug evolution runs on through *Scaphander*, *Haminea*, *Philine* and *Runcina*. In *Haminea* the shell is thin and bubble-like, in *Philine* it is transparent and quite internal, and in *Runcina* it is lost altogether.[10]

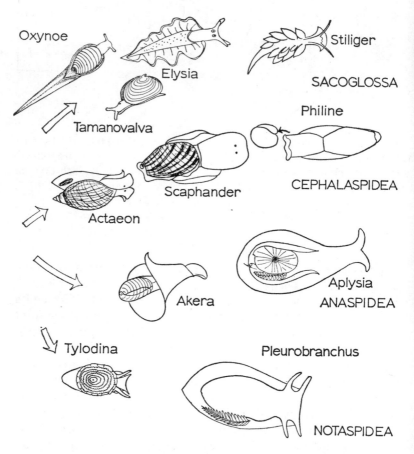

Figure 14 Some radiations and derivatives of shelled opisthobranchs: Sacoglossa, Cephalaspidea, Anaspidea, and Notaspidea

The burrowing apprenticeship and the streamlining of the body have set the way for more ambitious developments. The gastropod is now relieved of its cumbersome visceral coil and heavy shell, and can develop either the flat slug-like body of a nudibranch, or can become water-borne and swim. The Anaspidea or Aplysiomorpha are rather primitive opisthobranchs. They are not burrowers but have a plump high-built body, narrowing at the anterior end (Fig. 14). In *Aplysia* the vestigial shell is a small chitinous plate, with only a hint of spiral coiling. It is hidden in the roof of integument that reaches across the small mantle cavity and gill on the right side. Thin side-flaps of the foot called parapodia can close over all these structures like a jacket, to meet in the dorsal midline. In *Notarchus* these flaps are permanently closed, with only a small orifice admitting and expelling water from a sub-parapodial space.[204] In tropical *Dolabrifera* and *Dolabella*, the parapodia are also closed and the sole is broad and flat, adhering firmly to the rocky substrate.

In the Notaspidea, or pleurobranchoid slugs – such as *Pleurobranchus* and *Oscanius* (Fig. 14) – the body is much flatter, the mantle cavity has quite disappeared, and there are no parapodia. The ctenidium is left naked, sheltered only by the margin of the mantle skirt on the right side. The shell – as in *Aplysia* – is usually embedded in the dorsal integument. It is, however, sometimes external and unexpectedly large, for example in the saucer-shaped shell perched on the top of the fleshy body of *Umbraculum*, or the limpet shell of *Tylodina*.

In the true nudibranchs the shell, mantle cavity and gill are finally lost in the adult. Henceforward we shall see the lavish evolution of the naked upper surface, with all those functions that devolve upon it – sensory, respiratory, defensive and camouflaging. A good example of an early nudibranch is the British *Tritonia*, a member of the group of Dendronotacea. These slugs are distinguished by the branched or foliose processes often put out from the dorsal surface, and concerned primarily with respiration. The anus still lies on the right side. The eyes are vestigial, having become so in the earliest opisthobranchs. The osphradium is also lost, being replaced by a pair of new olfactory head tentacles, the rhinophores, already to be seen in the shelled opisthobranchs. As well, the whole dorsal surface is highly sensitive to tactile stimuli. Finally, in the Doridacea typified by *Archidoris*, *Goniodoris* and *Jorunna*, the anus has moved to the dorsal mid-line, and there is every appearance of bilateral symmetry save for the genital opening still on the right side. Around the anus appear a ring of five to nine pinnate secondary gills, which are sometimes retractile.

A different mode of respiration is used by the third group of nudibranchs, the Aeolidiacea. The integument is produced into slender club-shaped appendages called cerata. These either lie separately in several rows, or may be clustered together in tufts and palmate bunches. They contain blood from the haemocoele and are invaded also by tubular branches of the digestive gland which – as evolution proceeds – becomes deployed upon the dorsal surface.

The small slugs of the order Sacoglossa, which feed suctorially on green algae, are unrelated to other nudibranchs. Their whole dorsal surface may be respiratory, as in *Elysia* (Fig. 14) and *Limapontia*, while in *Hermaea* and *Stiliger* cerata may be developed, as in aeoliids. In *Elysia* and *Tridachia*, the flattened margin of the body is coloured by special zooxanthellae which multiply there.

The colours of nudibranchs are nearly always beautiful; and it is sometimes difficult with the animal away from its background to tell whether they are really warning or camouflaging. Many of the aeoliids are famed for their conspicuous reds, yellows and pinks upon a white skin. Aeoliids owe their immunity from enemies to the habit of feeding on coelenterates and storing the stinging cells of these in 'cnidosacs' in the cerata (p. 112). In some species, however, bright colours may offer a cryptic resemblance to a coelenterate background; in this way *Doto cornata*, with whitish cerata, is able to hide among the tentacles and sporosacs of the hydroid *Clava multicornis*. *Aeolidia papillosa* – with greyish-brown cerata – rests unnoticed beside the tentacles of its anemone prey, *Anemonia sulcata*. *Calma glaucoides* with silvery grey cerata likewise conceals itself against the masses of blenny eggs on which it feeds. In the pinkish *Tritonia plebeia* the branched marginal papillae resemble the half-expanded tentacle crowns of the polyps of *Alcyonium* colonies. The brilliant reds and oranges of many dorids must be primarily warning colours, as are the black and orange longitudinal stripes of *Pleurophyllidia*. In *Rostanga rufescens*, however, brick red has a camouflage value seen against the red sponge *Microciona*. The herbivorous aplysioids rely mainly on camouflage colours, such as mottled greens, browns and yellows, together with the secretion of an offensive purple dye. The carnivorous pleurobranchids on the other hand develop bright reds and yellows, or reds against milk-white, which may be taken to be warning colours, the animal producing a distasteful acid secretion.

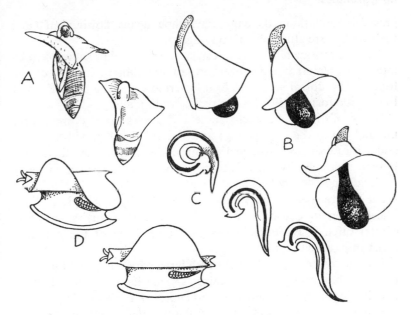

Figure 15 Some modes of gastropod swimming
(A) The prosobranch *Olivella tehuelchana*
(B) *Akera bullata,* a shelled aplysiomorph
(C) The lateral-bending dorid *Nembrotha*
(D) The pleurobranchoid *Oscanius*

Nearly every group of opisthobranchs has produced a swimming offshoot: flattening and streamlining are preadaptations as useful to a water-borne mollusc as to a bottom dweller. The simplest types of swimming are found in the Dendronotacea, where – as in *Dendronotus* and *Scyllaea* – the foot is reduced to a narrow groove, and the animal flexes its body from side to side to an angle of 45°. In *Dendronotus* a screw-like spiral flexure may run along the body. Appendages from the dorsal surface are useful in swimming, and *Scyllaea* progresses with its back downwards, from the weight of its tentacles and dorsal papillae. It must be an ungraceful swimmer, being said to bear a grotesque resemblance to a four-legged animal with ears, such as a Skye Terrier! The large plump *Melibe* and *Tethys* swim back upwards by rows of leaf-like dorsal paddles. They have a cowl-like cephalic hood which they toss from side to side to snare swimming crustacea. *Phyllirhoe* and *Cephalopyge* are built to quite a different plan, being small, transparent and laterally compressed, with the foot entirely lost. They are permanently planktonic, swimming with graceful undulations (the name Phyllirhoe means 'flowing leaf') and their bodies are studded with tiny light organs.

The pleurobranchid *Oscanius* can swim with a rolling movement by alternate up and down movements of the thin extensions of the foot.[293] The dorsal surface is held lowermost and the projections of the mantle form a bilge keel to reduce rolling and yawing.

Many of the nudibranchs 'take off' and swim by progressive undulations of the whole body; or by lateral bending, with the head and tail rapidly approximated at one side and then the other. As a human swimmer uses his extremities in a breast stroke, so many aeoloiids such as *Aeolidiella alba* and *Coryphella cyanara* stroke the water with the cerata. The Glaucidae are a family of pelagic aeoliids, bluish-green in colour, and spreading out three pairs of bushy ceratal tufts. These contain gas-filled diverticula of the gut, that assist in passive floating.

It is the lower opisthobranchs with large parapodia that turn their bodies to best account in swimming. The sea-hare, *Aplysia punctata*, is a simple example; it can take short spurts off the ground by rapidly opening and closing the parapodial flaps. *Aplysia saltator* makes an essay in jet propulsion, keeping the parapodia closed and contracting them upon the water beneath which is spurted from a funnel-like opening. Still more expert is the slug *Akera bullata*, which has a bulloid appearance, but is a primitive shelled aplysioid. Here the parapodia are prolonged into a graceful cloak, springing from the sides of the foot and overlapping above and behind the back. The visercal mass with its shell hangs like a clapper inside. The parapodia open and close with a fine medusoid movement, and – borne up on its skirt – *Akera* rises in short spurts and swims gregariously, especially during the spawning season.[231] *Gasteropteron* is a true bulloid, not related to *Akera*, yet it has by parallel evolution acquired an exactly similar mode of 'medusoid' swimming.

The most active opisthobranchs are, however, the pteropods or sea-butterflies, light diaphanous forms that spend their whole life in the plankton. They are divided into two groups – the shelled and the naked – that resemble each other only in the way they swim. The two parapodia are drawn out into membranous wings, with narrow muscular bases. The shelled pteropods, or Thecosomata, are, generally, held to be derived from bulloids. The earliest family, the Limacinidae, are small molluscs never more than a few millimetres in diameter, with transparent spiral shells, sinistrally coiled as in bulloid larvae. The foot has a broad sole and bears an operculum, and from its sides are produced long parapodia, or 'wings', muscular at the bases and thin towards the tips. *Limacina* rows itself upwards in a broadly spiral course using these as oars (Fig. 16) and drops as a dead weight by holding them together motionless above the body. Further evolution has achieved the emancipation of the body from its gastropod shape. The Cavoliniidae have a calcified, non-spiral shell, with bilateral symmetry. Through the needle-like *Creseis*, the broader and compressed *Diacria*, the tripod *Euclio* to the plump inflated *Cavolinia*, every detail of the design facilitates rowing, equilibration and maintenance of vertical position with a minimum of muscular work.

The Cymbuliidae are larger pteropods, up to five centimetres across. They have no true shell but a chitinous *pseudoconcha* which forms a boat in which the whole animal lies. In *Cymbulia* a pair of broad-based wings are attached along the sides. They are flapped like a butterfly's wings with an up and down rowing movement; and they show also an undulation from the front backwards, which provides a forward thrust. In *Gleba* and *Corolla* the pseudoconcha is a wide saucer and the wings have grown together into a heart-shaped fringe which is moved gracefully in swimming like a skirt or the bell of a medusa.[219]

The second series of pteropods, the Gymnosomata (Fig. 16), swim much faster than the thecosomes, and are so much modified that we

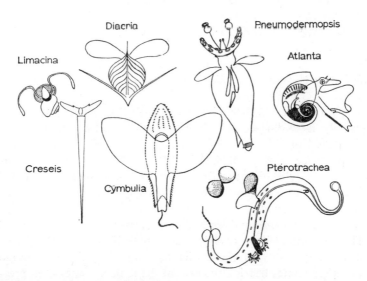

Figure 16 Heteropoda and Pteropoda

cannot be sure from which groups of opisthobranchs they have sprung. The body is torpedo-shaped, either blunt-tipped or pointed behind. There is never a shell, mantle cavity or true gill. Secondary gills may develop as flaps or fringes, on the right side or behind; or – as in the British *Clione* – respiration may be through the general body surface. The paired wings are narrowly attached beneath, behind the head, and rest horizontally, being moved dorsoventrally. The animal moves upwards or forwards by sculling with them, twisting them at the 'wrist' to allow a power stroke at both the upward and downward beat.[224]

Pulmonata

As compared with the prosobranchs and opisthobranchs, the pulmonates – though a successful and numerous group – are rather conservative in structure. The shell, however, shows many of the adaptations met with in prosobranchs, a land snail like *Helix pomatia,* the edible snail, being rather representative in shape, rounded or turbinate, with a wide aperture. A long anterior canal is never developed in pulmonate shells and there is only one family, the marine Amphibolidae, with an operculum. A frequent development, instead, is the presence of complex teeth and ridges, which to some extent guard the aperture against predators, as in ellobiids, vertiginids and many helicids. In the slender Clausiliidae there is in addition a loose sliding door known as a *clausilium* which fits into the grooves of the columella. During hibernation the higher pulmonates such as the Helicidae secrete an *epiphragm*, a temporary seal of dried mucus which may become very thick, and even – as in the Achatinidae – calcified.

There are several other widespread shell types. The first – found especially in small-sized primitive snails like the Endodontidae, and also in the Zonitidae – is flat and discoidal. These snails generally hide in crevices and under bark and ledges. Then there are the long spires of the Pupillidae and Cochlicopidae, or those of the Clausiliidae, tapered at either end, which may hang from dry walls or tree branches. Another type of long spire is the conical or fusiform shell developed in the Bulimulidae and their relatives.

The Pulmonata have also produced both limpets and slugs. The first are a speciality of the earlier and aquatic order Basommatophora, pulmonates bearing the eyes at the tentacle bases as in prosobranchs. The limpets of the marine Siphonariidae are the largest, and are entirely intertidal. They parallel the true limpets in their ecology, and on some tropical shores may replace them completely. Except for a shell sinus at the right side over the pallial opening, they are often difficult to distinguish externally from Patellidae. The minute *Otina otis*, living in high tidal crevices on western British shores, is a pulmonate limpet with a shell like an unperforated *Haliotis*.[221] In fresh water there are two entire families of limpet-like pulmonates, quite unrelated: the Latiidae, a primitive group living in Australasia, and the Old World Ancylidae, found both in lakes and in fast streams. In addition the Planorbidae have a limpet form in *Patelloplanorbis*, almost indistinguishable externally from a *Calyptraea*; and the Lymnaeidae have a peculiar limpet in the genus *Lanx*, from North America.

Pulmonate slugs belong to the terrestrial order Stylommatophora, with the eyes carried at the tips of one pair of tentacles. Almost every superfamily has at some stage contributed a specialized shell-less line. As well as economy in calcium, the slug habit has many structural advantages: these molluscs can glide through narrow spaces, or – like *Testacella* – burrow actively for animal prey which can then be swallowed into a distensible body. Pulmonate slugs are generally higher built and more slender than opisthobranchs, sometimes dorsally keeled. Here the body has been reorganized without loss of torsion; the mantle cavity, which serves as a lung, generally opens anteriorly on the right side. The mantle forms a fleshy saddle in which a vestigial shell plate, or granules of calcium carbonate, are usually buried. Some slugs – such as *Testacella*, *Daudebardia* and *Schizoglossa* – still carry a small limpet-like shell over the mantle. In the Testacellidae both shell and mantle are pushed back to the broader posterior end, and the front of the body is smoothly tapered for insertion in the soil.

An odd family of slugs, classed by some with the opisthobranchs, is the Onchidiidae, with one species, *Oncidiella celtica*, in Britain. These are marine and intertidal and have a posterior lung opening in the ventral mid-line behind the anus. The body is flattened with a thick warty integument, and the onchidiids are in some ways like naked limpets, resisting desiccation fairly well and making long journeys exposed to warm air. In some species the back is studded with small tentacles bearing eyes, and most forms have developed special repugnatorial skin glands with a defensive secretion. The colour is dull and inconspicuous, usually black, or mottled grey or brown.[128]

The terrestrial slugs of the family Athoracophoridae deserve special mention. Here the lung has lost its blood vessels and serves merely as a small vestibule with its wall produced into fine tubules that ramify in the underlying blood spaces. Though some suspect these canals to be glandular, critical experiments may yet confirm that these slugs are unique among molluscs in having developed respiratory tracheae.

The aberrant pulmonate slugs of the Vaginulidae, terrestrial in tropical America, Asia and Africa, have entirely lost the pallial cavity and respire by the moist integument.

In the land pulmonates, the shell is generally thin, and often fragile. Calcium may frequently be in short supply, and repair materials, if not provided with the food, may be drawn from elsewhere in the shell. Wagge[301] has made an important study of shell repair in *Helix*, showing the role of amoebocytic cells in lime transport. Calcium carbonate from the diet is stored in protein spheres in special lime cells of the digestive gland, whence it may be released to the lumen from time to time, for the regulation of the pH of the gut contents. Amoebocytes and alkaline phosphatase are both active in the digestive gland cells and at the site of shell secretion. The amoebocytes have access both to the lime stored in the digestive gland, and that laid down in previously secreted parts of the shell. No lime is stored at the secreting edge of the mantle. At the broken part of a shell, amoebocytes densely collect, arranging themselves in a sheet to form a continuous organic membrane, which is then calcified. The mantle epithelium plays little part either in forming the membrane or in laying down the deposit of calcium carbonate.

3 Burrowing and jetting —
bivalves, cephalopods and others

The gastropods have shown us a prolific variety in organization. The lamellibranchs and cephalopods – different though they are from each other – have concentrated on fewer structural patterns. Both attain great success by high specialization; and each class is very resourceful in the different ways its standard theme is given expression.

Lamellibranchs are much more sedentary than gastropods, though in most of them the foot is still well developed. Very few however crawl over the substrate in the primitive molluscan way. Many species burrow into soft sand and mud, or even bore into rock and wood. A large number are permanently anchored to the ground, and among these the foot is usually reduced and sometimes quite lost. Even here evolution is not exhausted: some of the least mobile of lamellibranchs have produced descendants that have broken free again to become swimmers, moving by expelling water on closing the valves of the shell.

In this chapter we shall give most attention to those bivalves where the shape of the body is less modified. Here the two symmetrical shell valves are drawn together by two equal adductor muscles, the one anterior and the other posterior. When these are relaxed the shell is opened by the elasticity of the *ligament*, which is a new structure peculiar to lamellibranchs. We may think of the mantle as originally a tent covering the whole body of the early mollusc. It soon became slit in front and behind so as to leave only a short connecting ridge in the dorsal line. Shell secretion was interrupted here and the ligament laid down instead, an elastic connecting strip continuous with the shell, but formed of uncalcified conchiolin, the organic substance of the shell. The ligament may lie slightly in front of, or more usually behind, the earliest point of the shell, the *umbo*; and it may be *external* (dorsal to the hinge) or *internal* (ventral to the hinge). The valves open by its elastic thrust when the adductors are relaxed, an external ligament being normally under tension, an internal under compression.[251, 295] Along the hinge line, the shell halves may develop teeth which interlock in various ways so as to prevent fore and aft displacement of the valves.

The adductor muscles probably arose from the enlargement and cross fusion of pallial muscles attaching the mantle to the shell. They now pass horizontally between the valves. Anterior and posterior pedal retractor muscles are also inserted on the shell near the adducts, and strike deeply into the body of the foot. The foot can protrude from between the valves and alter its shape by inflow of blood and contraction of its intrinsic muscles. Behind or at the base of the foot lies a gland from which many lamellibranchs – for part of their life at least – produce a *byssus*, a bundle of tough threads of tanned protein. In attached forms the byssus threads serve as mooring lines. In other species they are put out from time to time for temporary anchorage, or they may appear in the post-larval spat alone, the byssal gland becoming unimportant in later life.[235]

The mantle edge consists of three lobes (Fig. 8). The outer lobe secretes the outer calcareous (prismatic) layer of the shell; the middle lobe is sensory and the inner lobe or velum controls the flow of water. The thin *periostracum* is secreted by a groove between the outer and middle lobes, and the inner calcareous (*nacreous*) layer by the internal surface of the mantle. In earlier bivalves the mantle cavity is generally wide open; in *Nucula* for example the water current enters in front and passes out behind. In higher forms, especially those which burrow, the mantle margins are in some degree fused by the coalescence of one or more sets of lobes. The apertures for the water current move to the posterior end where the mantle edge is drawn out into tubular siphons, the inhalant one ventral and the exhalant dorsal. These elongate with increasing depth of burrowing and develop their own complex radial and longitudinal muscles. According to the number of mantle lobes involved, siphons may be naked and muscular (inner lobe only) (Tellinacea and Cardiacea); covered with periostracum by addition of the middle lobe (Mactracea and Myacea); or even encased in shell by the fusion of the outer lobe (*Cuspidaria*).[333]

The bivalve foot acts rhythmically in locomotion, not as in gastropods by waves of surface contraction, but by alternate lengthening and shortening. In protobranchs such as *Nucula* (Fig. 17) the sole is still a flat disc. But this is employed not for creeping as in gastropods but for thrusting ahead into the substratum like an anchor, after which the shell is drawn forward upon the fixed foot, as the pedal muscles shorten.

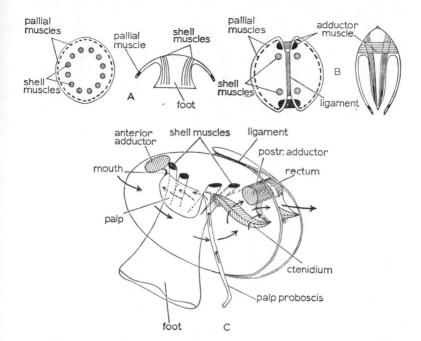

Figure 17 Derivation of the bivalve organization
(A) Before lateral compression but after appearance of pallial muscles; from above and in cross-section
(B) After compression, showing ligament and cross fusion of pallial muscles at either end; and in cross-section (*after Yonge*)
(C) Schematic early bivalve based largely on a nuculoid protobranch

Molluscan burrowing

Not only in bivalves but – as already seen – in gastropods too, hydraulic mechanisms have been developed for locomotion. In bivalves, and also in scaphopods – Trueman states – 'a double fluid-muscle system of internal (blood) and external (water) fluids is utilized'. With jet propulsion from the pallial cavity, the cephalopods are specialized to use only the external water. The haemocoelic spaces by contrast reach a maximum in the foot of those molluscs that burrow. Here the blood functions as a hydraulic organ for transferring the force of muscle contraction from one part of the body to another.

The fullest studies of bivalve locomotion have been made by Trueman, Brand and Davies, dealing with *Tellina, Macoma, Donax*

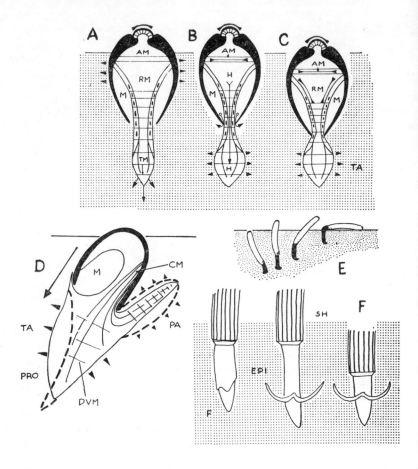

Figure 18 Molluscan burrowing

(*above*) Three stages of a generalized cycle of bivalve burrowing

(A) the foot probes downwards;

(B) the foot is maximally expanded and the siphons closed, water being expelled from the mantle cavity by shell adduction;

(C) the retractor muscles contract and the shell is drawn into the sand

(*left*) (D) Longitudinal section of a burrowing *Natica*, with the terminal anchor formed by the dilation of the propodium (from *Trueman*)

(E) Burrowing sequence of bivalve *Ensis*

(F) Burrowing of the scaphopod *Dentalium*, with the use of epipodium

AM, adductor muscles; CM, columellar muscle; DVM, dorsoventral muscles; EPI, epipodium; F, foot; H, haemocoele; M, mantle cavity; PA, penetration anchor; PRO, propodium; TM, transverse muscles; RM, retractor muscles; SH, shell; TA, terminal anchor and *Cardium*, based on modern techniques of recording pressure changes and body movements by transducers coupled to multi-channel pen recorders.

In *Tellina* (Fig. 18), the distal part of the foot forms a narrow blade when retracted; on expansion it broadens to a triangular muscular sheet, sharp-pointed and able, by pulsating movements, to be thrust into the sand. Following its probing and elongation the foot becomes dilated so as to anchor its outstretched end. The siphon tips, which keep contact with the surface, next close, to prevent water passing out during the following phase. This consists of adduction of the valves with maximum dilatation of the foot and ejection of water from the pedal gape. The retractor muscles of the foot are next strongly contracted, so as to pull the shell downwards into the sand. Finally, the adductors relax, the shell gape increases and a resting period precedes the next ensuing cycle.

Bivalve burrowing, by the rhythmic action of the foot, takes place in two stages: (i) *penetration*, by the probing of the foot with the shell firmly positioned by its valves pressing against the sand; (ii) *adduction* of the valves, followed immediately by pedal retraction.

The fluid/muscle system consists of two separate fluid-filled chambers, mantle cavity and haemocoele. During extension and probing with the foot, only haemocoelic blood is used, the foot-shape being altered by antagonism between its retractors and its transverse and protractor muscles. The foot now operates at a nearly constant volume, Keber's valve preventing outflow of blood, and pressure rising (in the razor shell, *Ensis*, to a maximum of 10 cm of water).

Adduction of the valves now affects both external and internal fluids together, producing in *Ensis* pressures pulses up to 120 cm water. This causes pedal dilatation, giving a new and secure anchorage, so that at pedal retraction the closed shell is drawn down. Pressure in the mantle cavity produces powerful water jets, that assist movement of the shell by loosening the adjacent sand. In bivalves with the mantle edges extensively fused this water (with the siphons closed) emerges in powerful spurts from the pedal aperture, having the same effects as water-jetting in sinking a pile. But in bivalves the jet is not directed down, which would loosen the foot, but upwards, from the pedal gape, to assist the passage of the shell. The burrowing bivalve takes advantage of quicksand effects: wet sand may be converted by agitation to its more permeable sol phase, as the foot is plunged down, setting again to a gel for the phase of firm anchorage.

As will be shown by cinephotography, the burrowing process

consists of a series of step-like movements into the substrate, the 'digging cycle' involving coordination of the muscular systems of the entire body.

The gastropods have only a minority of sand-burrowers, but these include some interesting specializations: Naticidae, Nassariidae and Olividae among prosobranchs, and the bullomorph families Philinidae and Hamineidae. *Natica* has long been noted for the massive dilation of its foot in locomotion, so great that it was suggested by Schiemenz in 1884 that the medium involved was not only the internal flow of blood, but also water taken in from outside through the spaces of an 'aquiferous system'. An expanded *Polinices josephinus* requires considerable tactile stimulus to withdraw, upon which small fountains of water have been observed from several pores round the edge of the propodium. The volume of water given off on contraction is here invariably two or three times greater than the whole shell volume. In other naticids it may approximate to, or slightly exceed, the shell volume; but with the evidence from *P. josephinus* it would still appear that sea-water can supplement blood flow from the pedal sinus in naticid locomotion. The mechanism of the aquiferous system, still largely unknown, would be of great interest to explore.

Brown has recently studied the expansion and retraction of the foot in the nassariid *Bullia*, convergently adapted with *Polinices* for life in soft sand. The contribution of an aquiferous system here appears to be negligible.

In the essentials of its burrowing, with massive extension of the foot, Trueman has compared the naticid snail in detail with the bivalves. *Natica* and *Polinices* form their terminal anchor by the dilation of the propodium or forward lobe of the foot, corresponding to the distal end of the bivalve foot. In bivalves the 'penetration' anchor is applied by the opening of the valves, but in naticids by the swelling of the mesopodium, prior to pedal extension.

The burrowing of scaphopods (*Dentalium*) shows essentially the same movements and sequence of activities as in Bivalvia. Digging consists of pedal protraction and retraction, with the application of shell and pedal anchors. *Dentalium*, however, lacks the water jets to loosen the sand and the high pressures in the pedal haemocoele. The strength of the pedal anchor is comparable to that of bivalves. But its probing force is relatively weak, since the shell anchorage comes from its weight rather than the pressure of open valves against the substrate.

Bivalve radiation

The locomotor repertoire described has made possible a wide adaptive range. The simplest of movements are probably the shallow burrowing of *Nucula*, using the flat soled foot as an anchor.

A number ' of the higher bivalves have a similar locomotion, burrowing shallowly or ploughing along the surface with a pointed muscular foot. In the freshwater mussels (*Unio* and *Anodonta*) the margins of the mantle are only slightly fused, to provide two posterior siphons, a wide inhalant and a narrow exhalant, scarcely projecting beyond the edge of the shell. The cockles (Cardiidae) are also surface or shallow burrowing bivalves (Fig. 19). The shell is plump and globular, the valves being thick and heavy, usually with prominent sculpture. In cockles and in some kinds of venus shells, strong ridges or sharp lamellae may keep the shell steady in soft sand. The rounded shape of some shells allows wave-rolling on the surface without damage. The valves of cockles make a tight fit, and the posterior third is often exposed. The siphons are short. The foot is long and narrow, rounded in cross section and of firm muscle. In some cockles, as in Cyprinacea and Trigoniidae, too, it can be flexed almost double and the tip inserted beneath the shell, lifting the animal off the ground in a powerful leap.

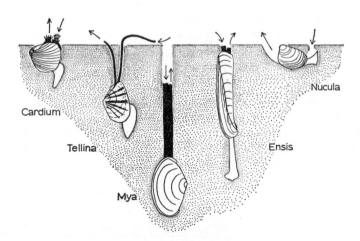

Figure 19 The burrowing habit in the Bivalvia

In deep-burrowing bivalves, the siphons elongate to provide the only effective contact with the world above (Fig. 19). They may acquire sense organs such as tactile papillae, light-sensitive spots or even elementary eyes. The siphons are inserted by special retractor muscles upon the interior of the shell, and the line of attachment or the mantle is here embayed to form a *pallial sinus*. Even in fossils, the depth of this sinus gives very reliable information about the length of the siphons, and thus the burrowing habits. Among the most mobile and accomplished burrowers are the Tellinacea, the Mactracea and some of the Veneracea.

With deeper burrowing, new specializations appear. In the more advanced lamellibranchs, particularly those of the order Adapedonta, the mantle edges are almost entirely fused, leaving only a small gape for the foot. The siphons are very long, generally lying parallel and surrounded with a common leathery sheath. These deep burrowers fall ecologically into two groups, the first typified by the fast-moving razor-shells (Solenacea) and the second much less mobile, containing the clams (Myacea) and the borers into hard substrata, like piddocks and shipworms (Adesmacea).

The shell of *Solen* or *Ensis* (Fig. 19) is long and thin, straight or slightly curved like a razor. It is beautifully adapted for slipping vertically or obliquely into the ground, and these bivalves will burrow as fast as one can dig. The shell is open at the front and hind ends, and its shape has been achieved by great elongation of the posterior half – the umbones are at the extreme anterior end. Fully half the mantle cavity is occupied by the piston-like foot, that can be protruded from the front of the shell, pointed and plunged into the sand. Once embedded it is made turgid with blood, and swollen to a thick bulb, while the shell and the animal are pulled after it. The ventral edges of the mantle are entirely fused, except for a minute 'fourth aperture' lying posteriorly towards the base of the siphons.

In the gapers and clams (Myacea) the shell is rectangular or ovoid (Fig. 19), in no way suited for fast burrowing. Bivalves such as *Mya* and *Panopea* live at a depth of up to 30 cm or more in compact sand or mud. The long siphons with their common leathery sheath can no longer be accommodated in the shell. The foot is small and the pedal opening narrow and easily closed. Rapid digging is in fact impossible, and if once dislodged from their deep stations the adult clams have little chance of re-embedding. The shell hinge is very simple and loses most of its teeth, since the valves need no longer be rigidly locked against desiccation or enemies. By the contraction of one or other adductor muscle, the valves can indeed be rocked in a horizontal plane on a fulcrum at the hinge. The front halves can thus be drawn apart, and in some clams like the American *Platyodon* hard substrata can be mechanically excavated by ridges on the shell. The burrow comes to form a mould of the animal, oval in section with dorsal and ventral ridges. Pallial hydrostatic pressure may be very important in the movements of these bivalves with closed mantle cavities. When the siphon tips and pedal gape have been closed, the siphons may themselves be extended by the contraction of the adductors against the pallial water. Or by building up water pressure, then closing and shortening the siphons, the valves may be opened against the wall of the burrow, with a consequent abrading action.

The boring habit is perfected in those bivalves that enter rock. There are already some Myacea such as *Saxicava* that nestle in crevices or existing holes, taking hold of the end of the burrow with the plug-like foot or the byssus, and abrading it with ridges on the shell. In the true rock-borers or piddocks (Pholadidae)[261] (Fig. 20) the mantle is closed save for a small pedal aperture in front. Here the valves gape widely and the foot can be extended and made turgid, to grip the end of the burrow like a sucker. The ligament in the piddocks is much reduced (Adesmacea). The valves are joined mainly by the

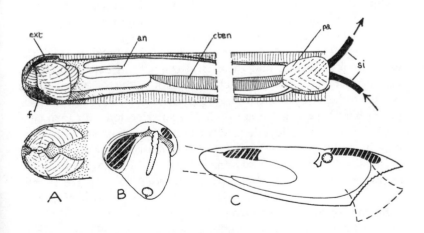

Figure 20 The Adesmacea: boring Bivalvia *Teredo* in its tube, with shell valves (A) dorsally viewed, and (B) left interior, compared with *Pholas* (C) an, anus; cten, gill; ext, external muscle; f, foot; pa, pallet; si, siphons

adductors, which can rock them transversely on the hinge fulcrum. Part of the anterior adductor has now spread outside the shell, and unites the valves dorsally to the hinge and in front of the umbones. In some species this exposed muscle is protected by one or more accessory plates of shell. The anterior and posterior adductors now serve respectively to open the valves and to draw them apart in front. On the anterior half of each valve the sculpture is sharp and abrasive, with teeth like a file. As the foot takes its grip the shell is hauled forward to the extreme end of the burrow. As the valves open, the shell is rotated alternately in either direction by special contractions of the pedal muscles. The burrow is thus made circular in section. The pholad shell is broadest in front, and the burrow is enlarged at its deep end so as to imprison the animal. In *Pholadidea* and *Martesia*, when boring ceases, the edges of the valves are approximated by further growth, the front of the shell becoming spherical, and the pedal gape closed so that the foot can no longer protrude.

Even in pholads the fused siphons are very long, and they may be protected by a horny or calcareous covering round their base. In the Teredinidae or shipworms (Fig. 20), which bore into wood, the siphons enlarge enormously at the expense of the rest of the body, and secrete the structureless calcareous layer which lines the burrow. The *Teredo* animal is limp and worm-like, consisting almost wholly of the conjoined siphons, the inhalant one containing an extension of the mantle cavity with most of the gill. The siphon tip is protected by shelly plates known as pallets, while at the anterior end the small visceral mass is covered by the true shell.[260] As in pholads, this forms the abrading tool. The hinge provides a rounded boss on which the valves rock sideways, and there is also a second ball joint at the ventral margin of the shell. The posterior adductor provides the power stroke, contracting to draw the valves apart in front, till their razor-sharp edges chisel the wood. The anterior adductor then contracts to draw them together. Between movements, the foot loosens its grip, and moves a little way round the burrow; the animal turns eventually through 180° and then reverses. In both pholads and shipworms, spoil from the burrow is carried through the mantle cavity by the pedal gape, and ejected from the siphons behind.[261]

So far we have passed over those numerous bivalves that live permanently attached at the surface. The reason is that most of these are far more modified than those that burrow. The majority of them form a fairly natural group, those – in the main – that used to be classed as Filibranchia and Pseudolamellibranchia (see p. 200). They are an older stock than most of the burrowers, but in shell form and habits they are some of the most specialized bivalves of all (Fig. 56). They include the true mussels (Mytilacea), the pearl oysters, wing shells and fan-mussels (Pteriacea), the scallops (Pectinacea), the saddle oysters (Anomiacea) and the true oysters (Ostreacea). The shell may be cemented to the substrate or occasionally lie free; but the majority – at some time in their life – anchor by the byssus. In the noah's ark shells and file shells, and rather less in *Mytilus*, the foot is still active. In most of the others it is vestigial or quite small. The mantle edges are little fused; there are usually no siphons and water generally enters round a wide circumference especially in the scallops, pearl oysters and true oysters.[329]

The mussels, Mytilidae, are less modified than most of this group. The byssus emerges in front from the ventral side and the foot lies at the anterior end. It can be protruded to prise the mussel free from the byssus and even to creep about. By encroachment from the foot and byssus, the anterior adductor muscle is reduced, and the posterior adductor enlarges at its expense. These changes are pushed ahead in later forms, until a single large adductor (the posterior) alone remains at the centre of the shell.

The evolution of bivalves with a single adductor (the Monomyaria) deeply involves the mantle cavity, and we shall postpone discussion of their adaptations until we can return to them with a fuller acquaintance with internal anatomy (see chapter 10).

Cephalopoda

The prime result of all cephalopod design has been to produce a swimming mollusc. The great majority of species move about by jet propulsion from the mantle cavity. We may regard all cephalopods, past and present, as having been able to swim, including in this class, by definition, all those early forms in which the body was first lightened by the incorporation of closed chambers in the apex of the shell. From one orthoconid nautiloid we have in fact fossilized impressions left in a soft substrate by the trailing of the conical shell as the animal alighted after jet swimming. There is good reason to think that the living *Nautilus* is, in its swimming, typical of shelled cephalopods in the past, and we may first consider the external structure of this form.

The shell of *Nautilus* is smooth, thin and light, forming a plane spiral, exogastrically coiled, that is, with the coil held aloft dorsally, and the mantle and body space lying below and opening in front

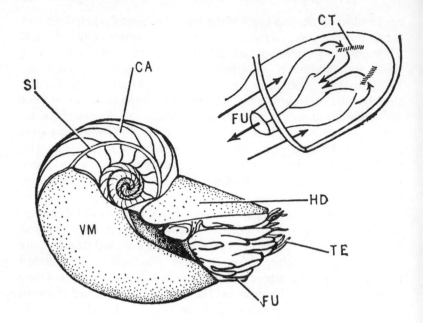

Figure 21 *Nautilus* in swimming position, with the shell chambers shown in sagittal section
(*above right*) ventral view showing the action of the funnel
CA, shell camera; CT, ctenidium; FU, funnel; HD, head shield; SI, siphuncle; TE, tentacles; VM, visceral mass

(Fig. 21). The shell is divided into some thirty compartments, increasing in size towards the most recent, which is the occupied body chamber. All the previous chambers are cut off by transverse septa. In *Nautilus pompileus*, the best-known living species, each new chamber overlaps and encloses the sides of the previous ones. They are filled with gas, resembling air but containing more nitrogen relative to oxygen. A shelly tube, the *siphuncle*, runs through the chambers to the apex but does not open into them. It carries a vascular siphon which is a narrow prolongation of the tip of the mantle. The gas enclosed in the chambers increases the animal's buoyancy, though its pressure was for long thought to be unable to be varied. The shape and extent of the body can, however, be adjusted within a considerable range by extension or retraction of the soft parts in relation to the mantle cavity. Living *Nautilus* is confined to the Indo-Pacific area. The three species live near the bottom in depths of up to 500 metres, but may have a considerable vertical range, coming into higher levels at night.

The *Nautilus* animal, while obviously related to squids and cuttle-fish, is nevertheless very different from all other recent cephalopods. The soft body is surrounded by the mantle and can be completely accommodated inside the mantle cavity in the final shell chamber. The tentacular crown of the foot, which surrounds the mouth, con-sists of two sets of lappets forming an inner and an outer circle. Their edges are fringed with small tentacles, some ninety in all, slender and annulated, and retractable into basal sheaths. They have no suckers, but are strongly adhesive. Above the head, against the rounded bulge of the shell, lies a fleshy hood which closes over the withdrawn animal. On the lower side lies the funnel, a modified part of the foot, formed of two separate halves that overlap in the middle line. Water passes into the mantle cavity round its whole edge, but the outward jet comes from the funnel alone. Sideways movements of the funnel are employed for changing direction. Water is forced out by the rapid retraction of the animal by its adductor muscles, and by the contraction of the funnel muscles themselves. At the same time all other exits are closed by contact of the body with the rim of the mantle. The mantle – being closely applied to the shell – is unable to contract in itself, as it does in all naked cephalopods.

Some recent investigations on living *Nautilus* have focused a new interest on this much-prized living relic. Denton and Gilpin-Brown have made a physiological and histological study of the buoyancy mechanism.[102] Dr Anna Bidder in 1960 made detailed observations of *Nautilus* in life.[70] Very sensitive to changes in light intensity, they remain quiescent by day. As illumination falls, the pre- and post-optic tentacle pairs emerge, and successively more tentacles un-sheathe, searching and grasping and passing particles to the mouth. Food is located by smell rather than sight. The animals in normal movement do not obviously or expertly steer round obstacles, depending rather on gently bumping and then altering course. They are slow swimmers with none of the visual acuity, living movement or colour change repertoire of more modern forms. 'They are odd . . . dumb and impersonal, a *long way* from other living cephalopods.'

Many of the early nautiloids were quite straight and concial. The external shell formed a long *orthocone*, divided up by septa and traversed by the siphuncle, which was sometimes very wide. Others were tightly or openly coiled, usually in plane spirals, but sometimes helical. We shall, in a later chapter, say much more about these and the evolution of the vast extinct cephalopod faunas,[65] but to understand properly the modern squids and cuttlefish we must refer first to the fossil belemnoids (Fig. 60). These had straight cigar-shaped shells, generally up to 12·5 or 15 cm (5 or 6 ins.) in length. From them arose all living cephalopods save *Nautilus*, and the belemnoids and modern cephalopods constitute together the sub-class Coleoidea.

The belemnoid shell was already internal, and differed from that of nautiloids in several ways. The chambered part – now called the *phragmocone* – was rather short, inserted behind into a much larger calcified *rostrum*, which forms the familiar cigar-shaped fossil. In front the phragmocone was produced into a chitinous shield, the *pro-ostracum*, to which the mantle muscles were attached. Such a shell gave at once support for the tissues, and the same combination of buoyancy and rigidity as we find in living squids.

Squids and cuttlefish, forming with the belemnoids the order Decapoda, have two long tentacular arms, able to be retracted into sheaths, and a circlet of eight short arms. The tentacles bear clusters of suckers at the tip, the short arms generally have them in several rows along the under surface. Each sucker is strengthened with a horny ring. In the octopuses and their near allies (order Octopoda) there are eight arms only, all long and tentacular; the suckers run right along and have no horny rings.

The modern cephalopod shell has lost many of its original parts (Fig. 59). In the pen of the squids the chambered phragmocone and the rostrum have disappeared altogether. The horny *gladius* and its *rachis*, or shaft, correspond to the pro-ostracum. The sepioid shell is more complete. In the well-known cuttle-bone of *Sepia* we have the persistent outer or uppermost side of the belemnoid phragmocone, now broad and flat, with the septa crowded together in the form of innumerable chalky laminae. There is a vestige of the rostrum in the pointed hinder end, but the pro-ostracum is missing. The smooth inner surface of the cuttle-bone represents the siphuncle so widely opened out that the lower wall of the shell has vanished. The small bathypelagic cuttlefish *Spirula*, three or four inches long, is unique among living forms in having a plane-coiled phragmocone, with an open spiral (i.e. a *gyrocone*), formed of about twenty-four chambers with simple concave septa and a siphuncle. The shell is almost surrounded by the tissues of the aboral fourth of the body, which bears rounded fins. The animal, buoyed up by the shell, hangs head downward, semi-vertically, and moves by short spurts of the funnel, catching small crustaceans and fish.

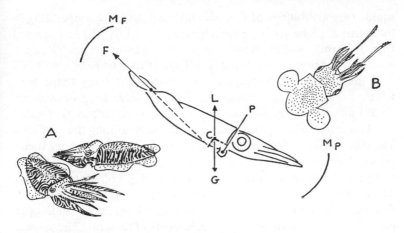

Figure 22 Dynamic lift in *Loligo* hovering in vertical plane
Benthic decapods (A) *Sepia*, (B) *Sepiola*

Most squids are long, and rounded in cross section (Fig. 22), whilst cuttlefish are shield-shaped and flat. The mantle is always thick and muscular, investing the lower aspect of the body, part of the sides, and opening in front. Its circular muscles can now contract freely, and this emancipation from the shell has allowed great improvements in locomotion. The funnel is now a complete conical tube, opening in the lower mid-line, and the power of the water jet is greatly increased. In modern decapods every principle making for quick propulsion and diminished water resistance has been beautifully exploited. Fishes alone can swim so fast or adeptly, and only in higher teleosts do we find again such a clean and single-purpose design for speed. The commonly observed squid *Loligo* – a foot or more in length – gives a good picture of a squid in action. The broad-based triangular fins serve as horizontal stabilizers; they are much used in hovering but are closed down in fast swimming. As well as the fast backward spurt, headward swimming is very efficient, with the reversal of the funnel direction, the retracted tentacles forming a wedge-shaped transverse prow. As in all fast cephalopods the side entrances of the mantle have developed fleshy valves and a *resisting apparatus* of cartilage studs and sockets. These serve as 'dome-fasteners' by which the mantle can be buttoned to the head on either side to prevent any passage of water except by the funnel.

The body form of a *Loligo* is depicted in Fig. 22, showing for these

c

squids the possibilities of dynamic lift, with the forces operating in the vertical plane to counteract gravity (G). Lift (L) is obtained from P (thrust from the funnel) and F (thrust from the fins). P and F operate a turning couple Mp and Mf about the centre of gravity C.

Among the modern cephalopods, habitat and body shape have widely radiated as new habitats and niches have been opened up, almost all of them by the emancipation from the bottom that speed and buoyancy have given to this class uniquely, among the molluscs. The coleoids have become adapted to swim, leap, walk, bury themselves, migrate up and down, even to fly. Ommastrephid squids in fact perform the only rocket or jet-propelled flight other than in man. The largest squids, like *Loligo*, round in transverse section with tail-fins triangular and streamlined for fast-swimming, are found in open oceanic waters, down to mesopelagic depths (1000 m). The largest, *Architeuthis princeps*, is the same length as the whale shark (17 m) with tentacles extended, though its body length is only a third of this total. *Chiroteuthis* has relatively the longest tentacles (see Fig. 25).

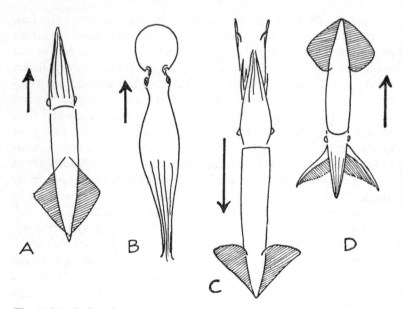

Figure 23 Swimming cephalopods showing direction of movement
(A) *Loligo* in forward attack
(B) *Octopus vulgaris* swimming
(C) *Gonatus* with hooked tentacles outstretched
(D) 'Flying' ommastrephid squid, with coriaceous arm membrane

The most powerful swimmers of all are the ommastrephid squids with the leading posterior edge like the profile of the mako shark and tunny. *Ommastrephes* and *Onycoteuthis* can take flight into the air when chased. Their stiff fins and coriaceous arm membranes are placed at opposite ends, like the fin-planes of a flying fish. But their lifting surface is much smaller, and it is apparent that most of the lift comes from continued water expulsion after take-off. Thus the Chilean giant squid *Dodiscus* can accelerate in air to a momentary maximum velocity of 14 knots!

Shown also in Fig. 23 is *Gonatus*, with its arms outstretched, carrying the highly modified hooked appendages.

In the shallow water benthic zone, in frequent contact with the bottom, the leading coleoids are octopods, sepiids and sepiolids. The cuttlefish, *Sepia*, is a leisured swimmer, hovering close to the bottom and feeding on shrimps and small fish stirred up from the sand by funnel jets. Along either side of the flat, shield-shaped body runs an undulant marginal fin, giving a slow backward or forward movement when the funnel is out of use. By shutting down on one or other side, the fins can be used for steering.

A related family of small benthic coleoids is the Sepiolidae (*Sepiola* and *Rossia*) with rounded bodies little more than 2·5 cm long. The fins are circular flaps attached at middle length, by which the animal scoops up the sand and nestles in it, on the watch for small crustaceans. The most dwarfed of all cuttlefish are *Idiosepius*, only 15 mm long and living in tide pools. The mantle develops a small sucker for temporary attachment to the fronds of green algae, such as *Ulva*.

The most highly modified of coleoids live in the deep bathypelagial zone, in permanent darkness. (*Cirrothauma* and *Cirroteuthis*, the deep-water octopods, p. 73).

Cephalopods, particularly the bottom-haunting sepioids, have brought the art of colour change and camouflage to a high pitch. Unlike the static patterns of opisthobranchs, the colours of cephalopods are under nervous control, unrivalled in delicacy in the animal kingdom. Total transformation of colour is possible in some species in less than two-thirds of a second. The skin has several types of chromatophore, small elastic bags filled with pigment, expanded by radial muscles and contracting by their own elastic power. The pigments are of the nature of *ommochromes*, derived from the amino-acid tryptophan. In sepioids the chromatophores form three layers, bright yellow near the surface, with a middle layer of orange red, and a deeper layer of brown chromatophores. In addition greens and blues may be produced as structural colours.

Sepia officinalis – as studied by Holmes[166] – has the finest colour repertoire. First, there are many colour changes that are concealing or *cryptic* in effect. The normal 'zebra' pattern of dark transverse stripes (Fig. 22) displayed when swimming, can be changed to a pale mottled brown and grey when lurking over a sandy bottom. On a white or pale background all the chromatophores contract to cause total pallor. *Sepia* can also produce a creditable imitation of a chequered black and white surface; such ability is probably of advantage on a ground scattered with large stones or white pebbles. On disturbance or 'anger' there follow vivid changes of colour giving a 'terrorizing' conspicuousness. Large black spots like eyes appear on the dorsal surface; or the body may momentarily become pale, except for a pair of black stripes flickering along the back, or alternating in rapid succession with the zebra pattern. When harder pressed, a flattened posture may be assumed, by depressing and broadening the body. The eyes become more conspicuous by bulging the irises, and the body colour alternates between general pallor and deep black spots. During hunting, brilliant colour waves may sweep over the back and arms, and a heightened impression of movement is created. The male *Sepia* also produces a sexual display, swimming alongside the female, spreading the broad lateral arms and making vivid play with the dark stripes.

Almost all cephalopods except *Nautilus* possess an ink sac opening just inside the anus, releasing a cloud of melanoid pigment to act as a protective smoke screen. Ink-discharge, combined with colour change, is also the basis of a neat deceptive trick. A small sepiolid squid will react to danger by first darkening, then emitting a compact cloud of ink to form a roughly shaped decoy, then immediately blanching and darting away unnoticed.

Buoyancy

The cephalopods, like the bony fish, have solved the buoyancy problem by the inclusion in the body of a gas space. An early achievement among the nautiloids and ammonoids, and basically unchanged with the chambered cephalopod shell since the upper Cambrian is the enclosure of gas at or below atmospheric pressure in the shell chambers, through the tubular siphuncle. In the living *Sepia*, *Spirula* and *Nautilus* gas is contained at lower than atmospheric pressures. As each chamber is formed, water is pumped out of it by the siphuncular membrane. *Sepia* can vary and control its buoyancy by altering the balance between the hydrostatic pressure of the sea and the osmotic force operating between the liquid in the cuttlebone and in the blood.

The cuttlefish shell then is comparable to a fish's swim-bladder, in matching the external hydrostatic pressure where the animal lives. The cuttlefish has a diurnal up-and-down cycle, burying itself in the sand by day, and rising to hunt by night; and there is a correlated cycle of buoyancy. The narrow spaces between the shell trabeculae, i.e. the much-modified shell camerae, contain gas, never under high pressure even in deep water. Fig. 24 shows the basic hydrostatic mechanism of cephalopods, illustrated by the cuttle-bone. Withdrawal of salt ions from the fluid immediately in contact with the

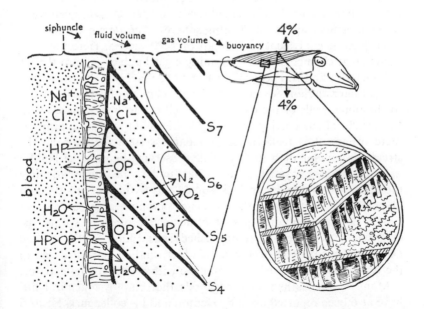

Figure 24 The basic hydrostatic mechanism of *Sepia*, showing effect of fluid volume, through gas volume, on buoyancy (*based on Denton and Gilpin Brown*)

siphuncular epithelium creates an osmotic pressure (OP) that counteracts the hydrostatic pressure (HP) of the blood and the surrounding water. Where HP exceeds OP, water is pumped into the cuttlebone, reducing the gas space. Excess of OP pumps water out, enlarging the gas space. The pressure in the gas space is determined by the partial pressures of gases diffusing from the fluid.

The vascularized inner surface of the flat shell corresponds with the remaining half of the widely 'opened' siphuncle of other shelled cephalopods. The shell is strong enough to withstand an external hydrostatic pressure of 20 atmospheres and the forces tending to 'implode' the gas spaces or crush the cuttle-bone are resisted by ribbon-like perpendicular bridges of calcified chitin passing between the successive septa.

Nautilus and *Spirula* shells have also been investigated by Denton and Gilpin Brown,[102, 103, 104] to whose important researches we owe almost all our modern knowledge of cephalopod buoyancy control. The engineering devices available to prevent implosion under high external pressures (thickening, vaulting, fragmentation of total gas volume, distribution of pressure along extended lines of compaction) have all been resorted to in cephalopods. Modern *Nautilus* can withstand pressures up to 50–70 atmospheres, *Spirula* up to 140–200 atmospheres, equivalent to a depth of 1400 metres. Here the shell has circular chambers, each containing a gas bubble no more than 1 millimetre in height. Packard[252] writes: 'The highly elaborate suture lines of later ammonoids marking the attachment of successive septa to the outer walls of the chambered shell are to be compared with the structural engineering of a late Gothic ceiling. What appears as mere ornament is really a rib and panel system for carrying the stresses impinging on a surface while preserving lightness of build.'

Cephalopod buoyancy control has undergone much adaptive radiation. Coastal and epipelagic squids – as we have seen – practise 'dynamic lift' (Fig. 22). Of the epipelagic octopods, *Argonauta* uses a simple device of holding gas at atmospheric pressure within the external 'shell'. Packard (unpublished) has found in the related *Ocythoe* the analogue of a swim-bladder, and an air-filled sac in the dorsal part of the mantle, complete with a duct to the exterior.

Many of the bathypelagic decapods that employ 'chemical lift' have also been reported upon by Denton and his colleagues. Neutral buoyancy is here achieved by using body fluids less dense than water. The bathyscaphoid squid *Galeoteuthis* has a fluid around the viscera, in a large cistern occupying about two-thirds the total body volume.

Its density is as low as 1·010, the cation being excretory ammonium, the only one suitable to keep the fluid lighter than, and isotonic with, sea water. Fluid with ammonium and chloride ions is contained by the Cranchiidae in the enormously enlarged coelomic space. *Histioteuthis*, *Octopoteuthis* and *Chiroteuthis* store such fluids in vacuoles that replace much of the musculature of the arms. The deeper-water Octopodids (*Japetella*, *Vitreledonella* and *Amphitretus*) with gelatinous tissues only sparsely muscular, appear to be neutrally buoyant.

Octopoda

In the living Octopoda all traces of the shell are finally lost. Only in the paper nautilus, *Argonauta* (Fig. 25), do we find a special case. Here the fragile 'shell' is a recent adaptive development, a neomorph formed of calcified conchiolin secreted by expansions of two of the arms. It occurs only in the female, where it forms a boat for carrying the egg-mass. The Octopoda are most typically represented in coastal waters by *Octopus* and *Eledone*. These are much more adapted to the bottom than the majority of decapods, though we must be careful not to generalize from them to the Octopoda as a whole. This order is rich in abyssal and bathypelagic forms, some of them strangely specialized and aberrant.

The body of an octopus is short and rounded. Fins are lacking, as in nearly all Octopoda, and there is little attempt at streamlining. Though *Octopus* moves fastest by swimming with the funnel, it also makes a much more versatile use of tentacles than do other cephalopods, and spends most of its life in intimate contact with the bottom. Short darting movements are made by gentle funnel spurts which increase in power until the animal leaves the ground, and swims horizontally with the tentacles trailing behind. On coming to rest the body may be pulled along nimbly, lightly supported on the tips of the suckered arms. The food, consisting of crabs and other slow crustaceans, is caught not by a lightning strike with a prehensile tentacle as in squids and cuttlefish, but with a swift pounce from above, with the arm circlet and its narrow web widely spread. As well as the visual powers common to nearly all cephalopods, *Octopus* has fine discriminatory powers in the sense organs of the tentacles: Wells and Wells (see p. 176) have demonstrated a well-developed 'chemotactile' sense, and the Octopodidae are pre-eminently the cephalopods of the sea-bottom.[310]

As reported by Young, *Argonauta* does not, like *Octopus*, appear to pursue its prey by sight, the eyes remaining motionless and the pupils of fixed diameter as objects pass near. But food is taken promptly when it happens to brush against the web of the first pair of arms, which is normally spread as a membrane over the shell. A regular response is made by the fourth arm which sweeps up the web and transfers it to the mouth. In *Tremoctopus* the arms and their webs form an enormous floating sheet, armed with symbiotic nematocysts acquired from the Portuguese man-o'-war, *Physalia*. Plankton is probably swept into the mouth with no discrimination, and the food-

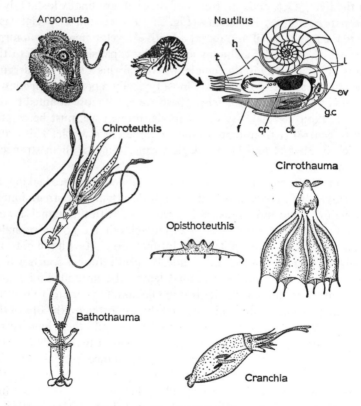

Figure 25   Some radiations in the Cephalopoda
Nautilus is shown in longitudinal section of shell, foot and pallial cavity
ct, ctenidia; cr, crop; f, funnel; gc, genital coelome; h, hood; l, 'liver'; ov, ovary; t, tentacles

detecting inferior frontal nervous system (see p. 174) is correspondingly reduced.

In the deep-water octopods of the super-family Cirroteuthacea, the web is very wide and supersedes the reduced funnel in swimming. The animal is bell-shaped, the web hanging like a skirt and the hump of the mantle carrying two broad fins. *Cirrothauma* and *Cirroteuthis* live at great depths and in total darkness. Uniquely among cephalopods, *Cirrothauma* has degenerate eyes and both genera probably rely chiefly on touch. The undersurfaces of the tentacles (inside the web) bear rows of tactile filaments on either side of the suckers, and these are said to gather fine particles of food which are then wiped off into the mouth in the manner of a holothurian.

In *Opisthoteuthis* (Fig. 25) the body is still more modified, flattened and circular like a jellyfish, fringed by the tips of eight tentacles. The mouth is at the centre of the web, and on the upper surface are two fins, a small funnel opening, and a pair of eyes. The last are very prominent, for the animal lives at no great depths but in the more shallow waters of the continental slope.

4 Mantle cavity and gills

Animals with delicate gills seldom expose them directly to the world outside; and in the molluscs one of the first distinctive features to evolve was the mantle cavity. This provides protection for the gills against injury or clogging with silt; and it allows an oriented water current to enter and pass out in a definite direction. The incoming current brings not only oxygen, but many other substances both useful and harmful to the mollusc. Particles of sediment, for example, must be carefully collected and rejected; chemical substances are constantly being tested by the osphradium; and in many molluscs microscopic current-borne food is brought in. Sperms, too, may travel into the mantle cavity of the female. Various products of the body are likewise extruded from the mantle cavity in the exhalant current, such as faeces, excretory matter, gametes and secretions, such as protective mucus, purple and ink. Indeed, from a mere cloaca with gills, the mantle cavity has evolved into the main vestibule and centre of commerce of the body. In those molluscs protected by a shell, it provides a space where the head and foot can be withdrawn to safety. As Graham has said, the mantle cavity is at once the strength and weakness of the molluscs. It offers many new mechanical problems; and once these are solved it opens up a new range of evolutionary possibilities. In the bivalves and some gastropods, for example, it provides the intricate ciliary feeding apparatus; and in cephalopods – equipped with powerful pallial muscles – it has become the chief locomotor organ.

Taking our primitive mollusc with two gills as the starting point, we find the mantle cavity lying posteriorly. It may extend along the sides of the foot by narrow grooves, and at either side of the rectum lie the paired organs (gills, osphradia and hypobranchial glands), known as the *pallial complex*. Or since – in typical molluscs – each gill is associated with an auricle and a coelomoduct, all these structures together may be referred to as the *palliopericardial complex*.

In the Amphineura[324] (Fig. 26) the mantle cavity is still primarily posterior. The more primitive chitons such as *Lepidopleurus* have the gills confined to this part of the body. The original pair has, however, increased to six or seven on either side, and the gill rows extend forward in deep channels between the side of the foot and the girdle. In this way the mantle cavity and gills reach as far forward as the head; higher chitons may have as many as seventy pairs of gills, the largest alongside the renal opening behind, and the size diminishing gradually in front and more abruptly behind. Each of these gills is a true ctenidium, with separate filaments, and the ciliation of each filament corresponds closely with that of gastropods. The curtain of gills divides the mantle cavity into two narrow chambers on either side: an inhalant chamber between the gill row and the skirt, and an exhalant chamber inside the gill row and against the side of the foot. Water may pass into the inhalant chamber at any point where the girdle is lifted from the ground, and the median exhalant current passes out behind the anus. The osphradium lies not at the entrance to the mantle cavity, but at the posterior end, though *Lepidopleurus* and some other chitons have additional sense organs at the base of each gill.

In both groups of worm-like Amphineura (Aplacophora) the mantle cavity is a small bell-like space at the posterior end. In *Chaetoderma* it can be rhythmically opened and closed. Simple though it is, this small mantle cavity – together with the radula – stamps the Aplacophora beyond doubt as early molluscs. *Chaetoderma* has two large plume-like gills, one at either side of the anus. In *Neomenia* these are even simpler, merely a circlet of small skin folds with blood from the haemocoele flowing directly into them.

In the Monoplacophora the five pairs of lateral gills in *Neopilina*, or the six in *Vema*, are unlike true ctenidia. They show none of the details of filaments or ciliary tracts so widely to be recognized elsewhere in molluscs, even in chitons where the multiplied gills are clearly ctenidial. It is more convenient to regard ctenidia as originally paired organs with a definite morphology situated in a posterior mantle cavity. Even where multiplied they supply no argument for metamerism. Nevertheless, both the Monoplacophora and the living *Nautilus*, with two sets of palliopericardial organs (p. 96), are unquestionably ancient. We may need to revise our assumption that the first respiratory organs were necessarily either ctenidial or posteriorly paired.

Prosobranchia

We have already seen in the Gastropoda (p. 23) how torsion has brought the mantle cavity to the front of the body and has reversed the topography of all the palliopericardial organs. We have suggested too that the forward-facing position of the mantle cavity must be of advantage to the adults as well as to the larva. Clean water is now drawn in from ahead of the animal and the pallial entrance is brought into a working relation with the anteroceptors, or sense organs of the head. With the chemosensitive osphradium at its base, the inhalant point of the mantle may be drawn out into a long siphon, and this may be employed as a movable nostril and a forward-seeking exploratory organ.

Torsion is brought about in the modern gastropod by the asymmetrical development of the shell muscles in the early larva. The single retractor muscle in existence when torsion begins has a right-sided insertion on the shell and sweeps over the gut to a leftward attachment to the foot. By its contraction the visceral mass is rotated dorsally and to the left (Fig. 5). The post-torsional right shell muscle is delayed in development until metamorphosis and then hyper-

---

Figure 26   Aplacophora and Polyplacophora
(A) A caudofoveate, *Falcidens*, in its burrow
(B) A hypothetical primitive placophoran with spiculiferous mantle
(C) A chiton, with spicules confined to girdle
(D) *Proneomenia*: schematic sections for comparison of shell, girdles and spiculiferous mantle
(E) Ctenidium of *Falcidens*
(F) *Chiton*, diagram of ventral aspect, with gill number reduced, showing inhalant and exhalant currents
(G) Horizontal section through three ctenidia in a chitonid mantle cavity, showing filament shape, location of lateral cilia and course of currents
(H) Diagram of a single ctenidium in side view
(I) *Chiton*: digestive, renal and circulatory systems
aff.v, afferent vessel; an, anus; aur, auricle; buc, buccal gland; cten, ctenidium; dig, digestive gland; eff.v, efferent vessel; ft, foot; gen, genital pore; gird, girdle; oes, oesophagus; osphr, osphradium; pall, mantle skirt; peric, pericardium; post.cten, posterior ctenidium; ren.ap, renal aperture; ren.org, renal organ; ren-per, reno-pericardial opening; stom, stomach; sug, sugar gland; vent, ventricle

trophies to form the principal (and in later gastropods the single) columellar muscle withdrawing the animal into the shell.[95]

The earliest shells generally classed with the Archaeogastropoda belong to the fossil Bellerophontacea. Here – as in *Bellerophon* and *Sinuites* – the shell is bilaterally symmetrical and coiled into a backward-directed plane spiral; it differs from that of any living archaeogastropod in showing the scars of two symmetrical retractor muscles. The palaeontologist Knight – who has published careful studies of these molluscs – holds that there is strong evidence that torsion had occurred in the Bellerophontacea, and that the mantle cavity faced forward. In the absence of asymmetry of the muscles, as suggested

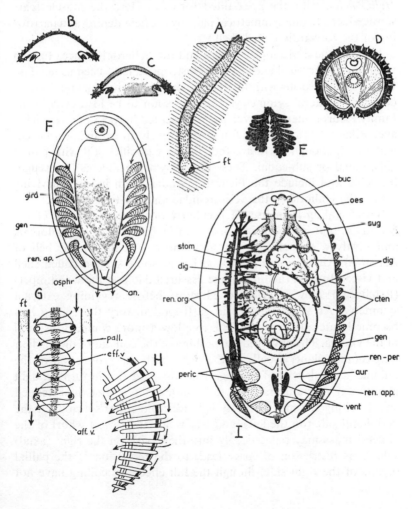

by the fossil scars, Knight suggests torsion could have taken place by the development of one muscle of the pair a little in advance of the other, by a mutation in the early veliger larva.[163]

Going back further than the Bellerophontacea, we find in the Cambrian shells that disclose no suggestion of torsion. These we must exclude by definition from the class Gastropoda and place in the class Monoplacophora. Such shells may be very variable in shape, and usually display several symmetrical pairs of muscle scars. A genus such as *Archaeophiala* in the Lower Cambrian has a cup-shaped shell with six to eight pairs of scars. This could have been succeeded by the taller horn-shaped shells of the Mid-Cambrian *Helicionella,* with the apex tilted forward. Here the muscle scars were reduced to one symmetrical pair, by the heightening and narrowing of the forwardly coiled shell.[162]

Between these Monoplacophora and the Bellerophontacea torsion must have intervened and the true Gastropoda have been ushered in. In the earliest forms with torsion the mantle cavity thus retained its complete bilateral symmetry. This was soon to be lost, even in the family Pleurotomariidae, which is represented in rocks of Ordovician age. Bilateral symmetry is found primitively in no living archaeogastropod, though there are four modern families still with paired gills, equal or sub-equal. These are known together as the Zeugobranchia and include the Pleurotomariidae, with its one surviving genus, the Haliotidae, the Scissurellidae and the Fissurellidae.

Spiral coiling is in itself a peculiarly deep-seated feature of the Mollusca, and is shown in some form by gastropods, cephalopods and bivalves, as well as – we now know – in the larval shell of *Neopilina.* Almost every gastropod has at some stage a coiled shell and visceral mass: the limpet-like Fissurellidae and the true limpets (Patellacea) show a brief coiled stage before regaining external symmetry. At an early stage in gastropod history the coiling had become bilaterally asymmetrical, to allow a more compact disposal of the viscera. This involved the pushing of the coil out of the median longitudinal plane, usually to the right, with the production of the familiar dextral twist of most gastropods. And this in turn brought about early rearrangements in the mantle cavity. After right-handed spiral coiling the mantle cavity is longer and more spacious at the peripheral side (on the left), and – as well – the lowest whorl of the visceral mass may bulge deeply into the cavity on the right (axial) side. This restriction of space leads to the reduction of the pallial organs of the right side, though the full effects of coiling have not yet appeared in the Zeugobranchia.[326] In *Pleurotomaria*, though spirally coiled, there are still two gills and hypobranchial glands, about equally large. In *Haliotis* and *Scissurella*, where the large shell muscle obtrudes against the mantle cavity on the right, the gill of this side is a little smaller. In the Fissurellidae, where coiling is lost, the gills are equal with resumption of bilateral symmetry.

In all these zeugobranchs the exhalant currents cross the gill to the mid-line of the mantle, where in various ways the shell and mantle cavity have been secondarily opened so that the exhalant current and the faeces may pass out more directly. In the Bellerophontacea the shell is notched by a median dorsal slit. In the living *Pleurotomaria* and *Scissurella* the body whorl of the shell and mantle are slit open by a long fissure as far back as the anus. In *Haliotis* there is a row of separate holes, the earlier ones sealing up as the shell grows. Some fissurellids, such as *Emarginula*, retain a slit in the front of the shell. In *Diodora* and the keyhole limpets, *Fissurella*, this closes to leave a small hole at the top of a volcano-shaped shell. In the slug-like *Scutus* the shell is reduced to a small shield and buried in the integument.

In the Zeugobranchia (Fig. 27A) each gill is attached to the floor of the mantle cavity by a suspensory membrane, and along the attached edge passes the efferent blood vessel to the auricle of the heart. At the free edge of the axis runs the afferent vessel from the wall of the mantle. Along either side of the axis, alternating at the two sides, runs a row of triangular gill filaments. Each is supported at one free margin by a chitinous skeletal rod, and consists of a double fold of thin integument. This encloses a narrow blood space communicating with the two blood vessels, and gaseous exchange takes place here.

The water current enters the mantle cavity ventrolaterally to either gill. Before reaching the gill, it is tested by the osphradium, then passes between the ctenidial filaments, and leaves the mantle cavity in the dorsal mid-line.[326]

Cilia are confined to the edges of the filaments and to a well-marked tract just behind the free edge overlying the skeletal rod. These last are the lateral cilia: they are responsible for passing water between the filaments and thus for creating the whole pallial current. Along the same edge of the filament runs a tract of frontal cilia; on the other edge lie the abfrontal cilia. These two tracts carry particles either to the tips of the filament or to the axis of the gill (Fig. 27B). Material may be rejected from the tips or carried anteriorly, and is compacted with mucus provided both by filamentar glands

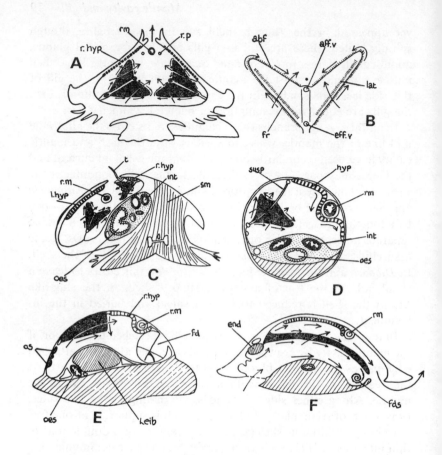

Figure 27 The gastropod mantle cavity in transverse section
(A) A zeugobranch gastropod, *Diodora*
(B) Gill filaments of *Diodora*
(C) *Haliotis*
(D) A higher archaeogastropod *Trochus*, with single bipectinate gill
(E) A monotocardian prosobranch, *Buccinum*, with a single mono-
pectinate gill
(F) A ciliary feeding monotocardian, *Crepidula*
abf, abfrontal cilia; aff.v, afferent vessel; eff.v, efferent vessel; end,
endostyle; f.d, female genital duct; fd.s, food string; fr, frontal cilia;
hyp, hypobranchial gland; rm, rectum; sl, pallial slit; s.m, shell muscle;
susp, suspensory membrane of gill

and by the large *hypobranchial glands*. These are thick folds of epithelium, lying on the mantle roof at either side of the rectum and mesially to the gills. A cross section is shown in Fig. 27D of the mantle cavity of a trochid, one of the higher archaeogastropods in which only the left gill remains. As in zeugobranchs this gill is bipectinate, with two rows of filaments. The right hypobranchial gland is lost, together with its gill, and the anus and both renal pores now open at the right side of the mantle cavity. The efferent region is thus no longer in the middle line, but has moved over to the right side. There are still two functional kidneys, the right one also passing genital products. The right auricle is all but lost, forming a small blind appendage and receiving no vessels.

The true limpets, Patellacea, are archaeogastropods specialized for the same mode of life as the chitons. The foot is a broad attachment disc, and the body appears bilaterally symmetrical. In the earliest family – the Acmeidae – the anterior mantle cavity still contains a single gill as in trochids. *Lottia* has developed in addition a series of adaptive 'gills' round the whole mantle skirt, except at the head end. These are not true ctenidia but form a row of simple flaps, hanging side by side from the mantle. The pallial cavity now reaches effectively right round the animal between the foot and the mantle skirt; respiratory water passes inwards between the pallial gills. In *Patina* (family Patellidae) the true gill is lost and the pallial gills function alone; finally in true *Patella* they form a complete circle extending over the head as well. Water can now enter and leave around the whole mantle edge. The limpets, being sedentary, have thus made an attempt at radial symmetry in the mantle cavity, though not in the rest of the body. What the chiton has achieved by multiplying the true ctenidia, the limpet has gained by substituting a ring of pallial gills.[326]

An archaeogastropod like *Trochus* is posed for the evolution of all the later prosobranchs, with a single gill and auricle. These fall into the two orders Mesogastropoda and Neogastropoda, and are sometimes known together as Monotocardia. Fig 27E shows a section of the mantle cavity of *Buccinum*. The gill is now monopectinate, that is, reduced to a single row of filaments. The frontal, abfrontal and lateral cilia are developed as before. The last-mentioned maintain the water current; the frontals and abfrontals collect particles which are rejected from the tips of the filaments towards the exhalant side of the mantle cavity. The hypobranchial gland is especially large and

secretes abundant and very viscous protective mucus. In the Murici-
dae and in some Volutacea a portion of this gland is responsible for
the well-known purple secretion, chemically a dibromoindigotin.
This is not known to have any protective function, and its role – if
any – is obscure. It is more probably an excretory substance.

The Monotocardia show various devices for regulating the direction of the ingoing and outgoing water currents. In higher mesogastropods and in almost all neogastropods the foremost part of the
shell aperture is produced into a long or short spout known as the
anterior canal. Through this runs a fold of the mantle forming an
incomplete tube which is the *inhalant siphon*. This may project well
in front of the advancing animal, or may be held erect like a median
hollow tentacle. It carries the effective opening of the mantle cavity
well in advance of the animal; and in carnivorous prosobranchs the
great development of this inhalant siphon, or 'movable nostril',
is related to the high development of the osphradium, which lies beside
the front of the gill just behind the base of the siphon, where it opens
into the mantle cavity.

The osphradium is the molluscan chemoreceptor which tests the
quality of the water entering the mantle cavity. It is extremely large
in carnivores, especially in the whelks and the cones, where it can
detect living and dead animal food at some distance away. The
simplest osphradia, as in Archaeogastropoda and lower Mesogastropoda, form merely a patch or a line of sensory cells. In later forms
the osphradium becomes bipectinate, like a small accessory gill, as
may be well seen in *Buccinum*. For *Neptunea antiqua* (12 cm long)
Yonge showed that the 240 osphradial filaments have a sensory area
of as much as 5 cm^2.[326] Particles are carried in between the filaments
by cilia, and at the free edge of each filament lies a narrow zone of
mucous cells and rejectory cilia. The passage of particles over the
filaments has led Hulbert and Yonge to suggest a further and perhaps more general function of the osphradium, that of a mechanoreceptor, to estimate the amount of sediment entering the mantle
cavity. In general, however, the largest osphradia are associated with
carnivorous life rather than with especially sedimented surroundings.
An exceptional carnivorous group are the pelagic heteropods, which
live in clean water; they have vestigial osphradia and detect their prey
with their powerful eyes.

In the South American, African and Indian freshwater snails of the
Pilidae (*Pila, Pomacea, Ampullarius*) respiration is amphibious. The
long inhalant siphon can be extended to much more than the length

of the body and reaches up to the surface of the stagnant oxygen-poor water where these prosobranchs live. The mantle cavity is partly partitioned by a fleshy fold, and develops a pulmonary chamber or vascularized lung at the left of the gill. By pulsation of the mantle, and inward and outward movements of the head and foot, the lung may be filled through the siphon with atmospheric air.

Life in turbid waters brings increasing sanitation problems. In many mesogastropods the gill filaments are prolonged and finger-shaped, with powerful rejectory cilia. Examples are found in *Aporrhais*, the Strombidae and the Xenophoridae. Such gills are pre-adapted for ciliary feeding, a device that has been fully exploited on soft substrata by the marine Turritellidae and the freshwater Viviparidae. The pallial organs become converted to function as collectors, not only of waste, but also of useful particles that can be ingested as food. This is essentially what has happened in the lamelli-branchs, and in a limited number of mesogastropods as well. One such family of ciliary feeders is in fact the New Zealand Struthiolariidae, the direct descendants of the Aporrhaidae, though unknown to Yonge when he foresaw this trend.[322] The gill receives an extra mucus supply from a glandular and ciliated tract, running along its axis in the path of the incoming current. Mucous strings are carried from here on to the frontal surface of the gill and thence to the tips of the filaments. Such a mucous tract was first described by Orton in *Crepidula*[243] by the rather unsuitable name of 'endostyle', by analogy with the ventral tract in the pharynx of early chordates. The glands on the gill filaments also contribute mucus for food-collecting, as possibly the hypobranchial gland as well. The main source of mucus apart from the endostyle is, however, a ciliated gutter running forward along the right side of the mantle cavity floor. Particles are thrown into this food groove from the tips of the gill filaments which dip down into it. They are then carried forward to the head, where the food groove debouches by a small spout behind the right tentacle. The proboscis and the radula are from time to time turned back to pluck strings of mucus-bound food from the opening of the groove.

The sand-dwelling wheel shells (Umboniidae), classed near the trochids, are the first archaeogastropods also shown to filter food with the ctenidium. The sessile vermiform prosobranchs (Vermetidae) have elongate gill filaments, but have not persevered with ciliary feeding, adopting instead the method of mucus-trapping by putting out long strings, or sometimes communal 'mess tables' from the pedal gland. These form efficient plankton traps, hauled in by the radula and ingested. The best-adapted ciliary feeders on hard substrata are the Calyptraeidae (*Calyptraea* and *Crepidula*), the Capulidae, and the Siliquariidae, vermiform shells closely convergent in form with the true Vermetidae.

Opisthobranchia

From the beginning of opisthobranch evolution the mantle cavity tends to move back along the right side to its posterior position as torsion is reduced. This involves the decreased importance of the gill, until – as we have seen – new methods of respiration by the adaptation of the dorsal body wall are provided in the nudibranchs. In primitive shelled opisthobranchs, the prosobranch ctenidium, with its thin ciliated filaments, is replaced by a different sort of pallial gill. This forms a fleshy horizontal fold, thrown into close-set numerous plicae; as represented in Cephalaspidea (bulloids) and in Anaspidea (aplysioids), it may be referred to as a *plicate gill*. It has no well-marked cilia, and the production of the water current has devolved upon two strips of pallial epithelium, with very powerful cilia, the dorsal and ventral raphe. These lie behind the gill, and sometimes extend into a spirally coiled caecum into which the mantle cavity – as in *Scaphander* and *Akera* – is drawn out behind. In the burrowing *Actaeon* this caecum is extraordinarily developed, being coiled independently around the whole visceral spire and reaching nearly to the apex of the shell. Its exact function is not known, but is probably to provide additional current drawing surface, and a larger flushing system for cleansing from sediment.

In the Notaspidea such as *Pleurobranchus* and *Umbraculum*, the mantle cavity is at last quite lost, but a conspicuous naked gill, neither a ctenidium nor plicate, still survives, under the right side of the mantle skirt.

In early opisthobranchs, however, such as the bulloids, that live on a sandy or muddy bottom, the hypobranchial gland and ciliary cleansing arrangements of the mantle cavity are still very important. The mantle cavity is frequently – as in *Scaphander* – drawn out posteriorly to the right of the animal into a spirally coiled caecum. This lies outside and quite separate from the rest of the body. In the burrowing *Actaeon* it is extraordinarily developed, coiling independently around the visceral spire, and reaching to the apex of the shell. The exact function of this caecum is not known – it possibly provides an accessory respiratory surface. The hypobranchial mucous gland extends right to the tip of it, and there are strong inward and outward ciliary currents that carry water through it, and serve as a flushing system for removing sediment.[138]

The thecosomatous pteropods, though they have lost the gill, have a broad shield-shaped hypobranchial gland and a very spacious mantle cavity. In the Limacinidae, at least, the pallial ciliary currents and the mucus from this gland serve to collect food particles: a food string is passed out from the exhalant side of the mantle cavity, as previously seen for example in *Crepidula*, and is periodically pulled into the mouth by the radula. In the Cavoliniidae, ciliated fields on the posterior sides of the wings collect food, which is carried to the mouth by ciliary tracts at the sides of the foot. In the Cymbuliidae these pedal tracts are elevated upon a spatulate 'proboscis', at the tip of which lies the mouth; ciliated fields on the wings are lost, and in *Gleba* the radula is lacking.[315]

In the aplysioids (Fig. 14) the mantle cavity has dwindled to a small triangular recess, at the middle of the right side. The fleshy gill partly projects from it, and there are two important sets of glands, both evidently protective in function. On the roof is a brownish-yellow gland – the homologue of the hypobranchial gland – secreting a deep purple substance (aplysiopurpurin). On the floor (Fig. 12B) discharge the grape-like clusters of the *opaline gland*, producing a colourless noxious secretion.[17]

Pulmonata

The pulmonates have followed a very different course. There is never a true gill, and though the shell may be lost, the mantle cavity nearly always keeps its anterior (post-torsional) position. Its roof is lined with an anastomosis of thin-walled blood vessels, and the cavity becomes air-filled, to act as a lung. The external opening is very small, a circular pneumostome on the right side, rhythmically expanding and contracting in land pulmonates. Nearby – outside the mantle cavity – open the anus and renal organs. The hypobranchial gland is lost, and the osphradium where present is usually extra-pallial; the pallial cavity is henceforward a closed lung and nothing more.

Such a lung does not confine the pulmonates to land: it is in fact an ideal organ both for aerial and aquatic respiration, and the majority of lower pulmonates, constituting the order Basommatophora, are truly amphibious.[220] Terrestrial evolution has occurred several times, and is seldom one-way. The Basommatophora fall into two series. Marine members – such as the siphonariid limpets, and the Amphibolidae – have become thoroughly re-adapted to submerged life. The mantle cavity is again filled with water and the osphradium is internal. A secondary intra-pallial gill, built up of narrow folds of epithelium, is developed, though never a ctenidium. The higher Basommatophora are dwellers in lakes, ponds and rivers; they have arrived there not from the sea or estuaries, but by varying degrees of re-adaptation from terrestrial life. The Succineidae – living amphibiously in marshes – are classed as primitive Stylommatophora; yet they are extremely similar in many ways to the Lymnaeidae, and some members of this family, such as *Lymnaea truncatula*, and *L. palustris*, are marsh dwellers with an air-filled lung, and live essentially out of water. Other lymnaeids, such as *Lymnaea stagnalis*, come to the surface regularly to fill the lung with air, and are said to drown if this is prevented. At the other extreme are deep-water species, such as *L. abyssicola*, which never surface. Intermediate species include perhaps the majority of Lymnaeidae, and also the Physidae, such as the British *Physa fontinalis*. In *Physa* and in *Lymnaea peregra* – a ubiquitous species tolerating many habitats – Hunter[171] has found several distinct physiological states in different populations. Individuals living near lake margins can occasionally come to the surface and take in air, as they do frequently under laboratory conditions. In others, further from land, the mantle cavity may contain a gas bubble that is possibly used as a physical gill; while in those in deepest water the cavity is permanently filled with water and respiration is entirely aquatic.

The Planorbidae and the Ancylidae are the most purely aquatic of pulmonate families. Here the pallial cavity is disused, and in Ancylidae it is almost lacking. Instead, a secondary external gill is developed by the enlargement of a pallial lobe, lying just outside the pneumostome. In the ancylids this lobe is served by the ordinary vascular circuits of the mantle, but at least in *Planorbis corneus*, a complete afferent and efferent circulation is developed. The external gill becomes elaborately pleated and folded, though it is never ciliated like a ctenidium. Several species of planorbids, such as the tiny *P. crista*, can tolerate very foul water; and *P. corneus* possesses haemoglobin serving as a respiratory aid and oxygen store in conditions of low oxygen pressure.[171]

Filter feeding by a mucus-trap mechanism has been recently described in the pulmonate limpet *Gadinalea nivea*, clustering immobile in large numbers, in caves and recesses with high water tur-

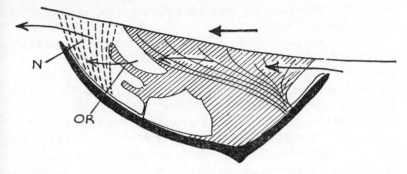

Figure 28 Gadinalea, with pallial secreted mucous net (N) and ora lappets (OR)

bulence. With the loss of the ciliation of the mantle cavity, food-collecting is now extra-pallial, relying on particles brought by water movement. As the animal clings upside down to the rock, the shell is lowered and a mucous screen secreted between the substrate and the mantle edge (see Fig. 28).[302] Phytoplankton lodged in this net is periodically eaten as the mucous screen is seized by the reduced radula. The prosobranch *Olivella biplicata*, burrowing shallowly in surf-swept sand, similarly uses a pair of mucus traps, held between the propodium and the head, to filter particles brought by back-wash of the waves.

Bivalvia

The most complex molluscan gills are found in the lamellibranchs. With some exceptions the bivalves are sedentary ciliary feeders, and the ciliary and mucous tracts of the mantle cavity play the central role in their life. In the evolution of their filtering organs the lamelli-branchs are paralleled by other ciliary feeders, such as polyzoans, brachiopods, tubicolous worms, Amphioxus and tunicates. The resemblances extend to the details of the lateral cilia which create the current, the frontal cilia which collect the food, and the tracts of mucus glands and cilia which transport and sort it. But the lamelli-branchs stand out alone in the beautiful elaboration of these mechan-isms, especially the sorting and straining cilia. Filtering is very efficient: particles down to 1μm in size can often be retained, and many species rely chiefly on the 'ultra-plankton' as food. The papers of Orton and Kellogg,[181] and more recently of Yonge,[316, 319] Atkins[55] and McGinitie,[198] have built up a lively picture of the working of the bivalve mantle cavity.

The gills are bipectinate (i.e. with two rows of filaments) and equal on either side. To understand their derivation from earlier gills, we must first describe one of the primitive Protobranchia, such as *Nucula*,[323] where the gills are relatively smallest (Fig. 29A). They lie behind the foot at the back of the mantle cavity, with the inner filaments from either gill touching in the mid-line. Such ctenidia, though larger, clearly recall those of early gastropods and their forerunners; though uniquely specialized in some characters, the protobranchs are in other ways a fine transitional group. In *Nucula* the filaments of each gill are triangular leaflets. In section the gills extend across the mantle cavity to form together an inverted W with short wide limbs.

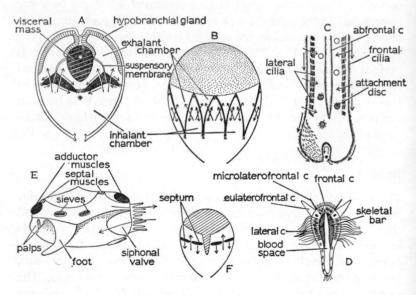

Figure 29   The bivalve mantle cavity and gill
(A) *Nucula*, in transverse section
(B) Ctenidial arrangement in filibranchs and eulamellibranchs (*from Yonge*)
(C) *Mytilus*, portion of a gill filament
(D) Section of a eulamellibranch gill filament (*after Atkins*)
(E, F) *Poromya*, a septibranch, left side view, and transverse section (*from Pelseneer*)

The inhalant chamber lies ventrally and the exhalant chamber above. Currents are carried between the filaments by the lateral cilia. Along the lower edge run the frontal cilia, on the upper edge the abfrontals. Their beat is so arranged as to carry particles towards the middle line, where the two inner filaments loosely interlock by cilia. Here the particles are rejected into the inhalant chamber. Above the gills, in the exhalant chamber, are found paired hypobranchial glands, absent in most later lamellibranchs. The anus and renal organs open into this chamber, and there is also – near the posterior adductor muscle – a pair of small osphradia.

The pallial water current enters in front and after passing through the gills goes out behind. The current serves for respiration, and the gill cilia, while rejecting particles, do not in *Nucula* collect food. On either side of the foot, lying in front of the gills, are the very broad labial palps, which are the largest of the pallial organs. On their inner faces they are lined by ciliated sorting ridges, and their posterior ends are produced into long grooved tentacles, known as the *palp proboscides*. These emerge from the shell and collect food particles from the soft substrate, which are carried in by cilia and sorted on the palps. Thus *Nucula* is not a ctenidial feeder; but the development of labial palps was – as Yonge has shown – perhaps a necessary stage in the evolution of filter feeding, first enabling the mouth to be raised above the ground.

Like some other primitive groups, the Protobranchia are very diversified.[323] In the Malletiidae (including *Yoldia* and *Malletia*) the gills are strongly fused by cilia in the mid-line. The interfilamentar spaces are reduced to rows of pores, and the gills acquire muscle fibres, functioning as a rhythmical pumping membrane drawing water from inhalant to exhalant chambers. Both inhalant and exhalant currents are posterior, passing through short siphons. In the Solenomyidae the shell is tubular and flexible, being only partly calcified. The mantle edges partly fuse and the foot passes through the anterior end like a muscular piston, extruding mucus-bound sediment from the mantle cavity. By forcible expulsion of water, darting and swimming movements are performed. The ctenidia are large and their currents collect food, the palps being small, without proboscides and are never protruded from the shell.

In the remaining bivalves the gills are more specialized. A history of this class could be written from their ctenidia, and their degree of specialization corresponds fairly well with the general evolutionary level. Above the prosobranchs the gill filaments elongate, then double back upon themselves. The first stage – the *filibranch* condition – is seen, for example, in *Arca, Mytilus* and *Anomia*. Each gill separately now forms a W in section, with long narrow limbs. Fig. 29c shows some points of anatomy. The gill axis lies at the middle angle of the W. The central limbs are the *descending lamellae*, the outer limbs the *ascending lamellae*, made up of the reflected distal parts of the filaments. Each V of the W forms a demibranch; there are thus an inner and an outer demibranch, each two lamellae thick, in either gill. In each demibranch the component filaments become attached to each other side to side. With filibranchs this is achieved simply by scattered discs of stiff cilia that interlock like hairbrushes; for this reason the filaments are easily pulled apart and the gill may take on a frayed appearance in dissection. In addition, connective tissue junctions run between the descending and ascending limbs and thus hold together the two lamellae of each demibranch.

In the gills of Ostreidae, Pectinidae and Pteriacea, which are described as *pseudolamellibranch*, the reflected distal tips of the filaments have coalesced laterally with the mantle, and mesially with the base of the foot (or further back as between the two gills). The gill has a greater cohesion than in filibranchs. Finally, in the *eulamellibranch* gill of the most advanced bivalves, the adjacent filaments are united by vascular cross connections, leaving narrow openings or *ostia* between them. The two lamellae of each demibranch become attached back to back in the same way. In most lamellibranchs the demibranch surface is without folds and flat, with the filaments all alike – the condition known as *homorhabdic*. In pseudolamellibranch gills and in those of some eulamellibranchs (Cardiacea and Myacea) the gill becomes more complex, with its surface plicate, or thrown into folds and grooves. In pseudolamellibranchs the filaments are of two sorts (*heterorhabdic*): about twenty *ordinary filaments* constitute each fold and a single enlarged *prinicpal filament* runs along the groove between two folds.

The cilia of a typical eulamellibranch filament are shown in Fig. 29D. Abfrontal cilia are no longer found, since the edge of the filament has been enclosed between the limbs of the V. These still persist however in filibranchs (Fig. 29C). Frontal cilia are very prominent along all that aspect of the filament that faces the mantle cavity, i.e. they cover the free face of the gill. The lateral cilia have the same position, just behind the frontal edge, as in gastropods. They pass the water current between the filaments, through the ostia, into the interlamellar space which leads above into the *suprabranchial cham-*

ber, carrying water backwards towards the exhalant siphon. Lying between the lateral cilia and the frontals is another set of cilia peculiar to the bivalves, the *laterofrontal* cilia.

A valuable study of the ultrastructure of the laterofrontal cirri by H. J. Moore, shows them to be of compound structure and terminally 'frayed' into branches, well capable of trapping ultramicroscopic particles (Fig. 30).[210]

These structures form triangular platelets, probably from the fusion of several separate cilia. They beat relatively slowly towards the middle of the frontal edge, at right angles, that is, to the frontal cilia. Each row of eulaterofrontal cilia forms a flexible comb, and the two rows bordering each ostium provide a sieve stretching right across the water space, and straining off particles for retention on the frontal surface of the gill.[55]

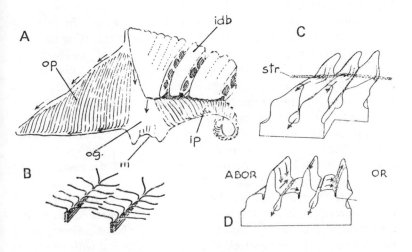

Figure 30 Gills and palps of *Venerupis pullastra* (*after Foster-Smith*)
(A) Insertion of the inner demibranch between the palps
(B) Detail of two laterofrontal cilia
(C) Acceptance by palp, with food string taken in orally
(D) Opening of palp ridges for entry to rejection tract
ABOR, aboral aspect; OR, oral aspect; idb, inner demibranch; ip, inner palp; m, mouth; og, oral groove; op, outer palp

Various tracts of cilia beat anteriorly and posteriorly along the margin of the gill, transporting either collected food to the labial palps (in oral grooves) or rejected particles backwards. Such currents run in grooves along the demibranch margin or dorsally at the inner or outer base of the gills. Particles are brought to them by the frontal cilia beating up or down the filaments.

The filtering power of the gill may be varied by altering the width of the ostia or the speed of the current; filtration is probably chiefly the function of the laterofrontal cilia. Some workers, especially McGinitie,[198] have held that particles are filtered and transported over the gill in a thin sheet or network of mucus, secreted partly by the filaments, partly by glands at the base of the gill. The mucous sheet has an extremely small mesh size and bacterial particles can be filtered and large molecular aggregates perhaps absorbed upon it. Thus clams can be fed for long periods on filtered meat extract, and in some experiments haemocyanin and haemoglobin were removed from the water in molecular form.

All authors would allow the importance of mucus in rejecting waste. Some, however, have questioned whether the sorting mechanisms of the gill could function, if particles were wholly embedded in a mucous sheet. Ciliary sorting is highly important in the bivalve gill, and Atkins has described the varied and ingenious methods, all contriving to exclude over-large or coarse particles from the oral grooves running to the palps and mouth.[55]

An important account of the functioning of the gill and palps has recently been given by Foster-Smith, working with *Mytilus*, *Cardium* and *Venerupis*.[125] In a bivalve opened for dissection, the mucous sheet on the gill generally ceases to form; but observations of intact specimens, through shell 'windows', show the vital role it plays in food-collecting. Bound particles are ultimately carried anteriorly along the gill margin, and sometimes also in the basal grooves, to the palps. Even at low concentrations, material is bound with mucus on the gill. The relation of gill to palps, and the fine structure of the laterofrontal cirri, are shown in Fig. 30.

The mucous sheet may be normally so fluid that the sorting cilia can project through it, or manipulate particles within it.

Particles excluded from the oral groove are thrown off the gill on to the sides of the foot or the mantle wall. The finer material carried in the grooves eventually arrives at the point where the demibranchs terminate between the bases of the labial palps.[316] Though smaller than in protobranchs the palps are very prominent, hanging as a pair of triangular flaps on either side of the mouth. Their outer sides are smooth; inside – where particles impinge – they are traversed by ciliated ridges and grooves. The palps form a sorting mechanism of a kind used repeatedly by molluscs, both in the mantle cavity and in the stomach. Heavier or coarser particles are carried into the grooves, where they are passed to the margin of the palp and rejected from its tip. Lighter material moves up the palp across the crests of the ridges to arrive at the mouth. Sorting is by size and weight, with little reference to quality, and coarse and unsuitable particles are at times found to enter the stomach.

In turbid waters the overspill from the gills and palps must be continuous. It is bound together with mucus and rejected from the mantle cavity as *pseudofaeces*; exit is by the pedal gape, or – in those bivalves with an extensively fused mantle and long siphons – pseudofaeces may travel posteriorly in a ciliated rejection groove running along the pallial suture to the base of the inhalant siphon, which expels them. The exhalant siphon passes only water, with renal products and true excreta; pseudofaeces are unable to reach it.

In some lamellibranchs part of the normal gill is lacking. The small, free and commensal bivalves of the Erycinacea have reduced or lost the outer demibranch, perhaps owing to the decreased gill surface ratio required with reduced body size. In the Teredinidae, where the gill is prolonged into the base of the inhalant siphon, only a modified inner demibranch survives. In the Anomalodesmata the outer demibranch is turned up dorsally and often loses its reflected lamella.

The same is true of many Tellinacea, and in this superfamily the gill surface is much reduced in comparison with the labial palps, which enlarge to approach the gill in size. All the Tellinacea are deposit feeders on the rich organic layer of the surface of the substratum, which they reach with the long inhalant siphon. Yonge has pointed to fundamental differences between Tellinacea and the majority of bivalves that feed by filtering suspended particles. The food deposit is coarser and heavier than plankton; it is sucked in with the inhalant current and by muscular action of the tip of the inhalant siphon which wanders freely over its available territory. Much of the entering deposit is probably drawn forward and thrown directly on to the surface of the palps; both here and in the stomach, sorting must be very rigorous. As compared with suspension feeders, the gills play a much smaller role in the collecting and grading of particles.[327]

The most extraordinarily modified lamellibranch mantle cavities are found in the small group Septibranchia (Fig. 29E, F), sometimes considered a separate order, and containing only three genera – *Poromya, Cetoconcha* and *Cuspidaria*. The ctenidia no longer exist as such, being converted into a horizontal muscular septum, spreading from the base of the foot to the mantle, and extending right back to the siphons. The septum is inserted on the shell by its own muscles, and completely divides the mantle cavity into ventral and dorsal chambers, communicating with the inhalant and exhalant siphons respectively. It is perforated in *Cuspidaria* by four or five pairs of ostia, in *Poromya* by two pairs of larger branchial sieves. It can be raised and lowered to form a pump, driving water intermittently through the openings from ventral to dorsal chamber. In *Cuspidaria* the septum develops peculiar striated muscle fibres; in *Poromya* it is much more delicate, and this genus leads on from *Verticordia*, one of the order Anomalodesmata with more normal but already very muscular gills. Septibranchs live in very deep water, burrowing shallowly in mud and ooze. They are no longer ciliary feeders but scavengers, ingesting the whole bodies or fragments of dead or moribund crustaceans and other small animals. As well as the gills, the labial palps are much reduced and retain no sorting function.[318] Reid and Reid[272] have given a good account of carnivorous feeding in *Cuspidaria*.

Cephalopoda

Here – as we have seen – a very powerful pallial current is created not by cilia but by muscular contractions of the funnel, in *Nautilus*, and the mantle in other living forms. This provides the motive power for jet propulsion, and the increased flow of water through the mantle current also meets the respiratory demands of higher metabolism and greater activity. In a squid, cuttlefish or octopus, the mantle cavity is a deep space enclosing the lower surface and sides of the body. It contains a pair of large bipectinate gills suspended one at either side of the rectum (Fig. 31B). Increased size and the strong pumping force mantle have necessitated great modifications in the structure of the gills. They are suspended by their *afferent* edges, not efferent as in gastropods, and the water current is driven from afferent to efferent side. The filaments are firm and fleshy, no longer ciliated, and thrown into primary and secondary folds to increase the respiratory surface. In modern cephalopods – not

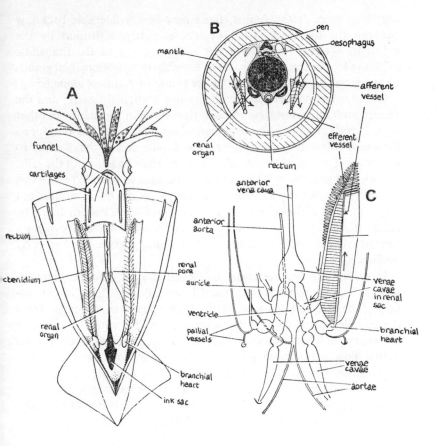

Figure 31 Pallial and circulatory organs of *Loligo*
(A) Ventral view
(B) Cross section
(C) Heart, venous and branchial circulation

Nautilus – the flow of blood through the gills is increased by the development of fine capillaries within the filaments. In addition, the blood flow through each gill is assisted by a pulsatile *accessory branchial heart*, lying at the base of the gill on the course of the afferent ctenidial vessel, and placed in an annexe of the pericardium. Pallial contractions drive water between the gill filaments at great pressure. As well as the normal afferent to efferent flow (Fig. 31B) outwards towards the funnel, there may be considerable backflow between the filaments. This pressure is partly withstood by the chitinous skeletal rods along the afferent edges of the filaments. With the loss of the ciliary tracts of the gill, the hypobranchial glands disappear as well. The osphradia are found in *Nautilus* alone.[326]

Nautilus has neither gill capillaries nor branchial hearts; and the funnel contractions are less efficient than those of the mantle in other cephalopods. Lacking these improvements, *Nautilus* has developed increased respiratory efficiency by the duplication of the ctenidia to two pairs. These gills are attached only by their bases. With the gills, the auricles and the renal organs also increase to four. Whether this duplication was ever widespread in cephalopods, or in molluscs at large, or a special occurrence in a few genera, we cannot tell. The recent discovery of *Neopilina* – with five pairs of these structures – suggests that this may have been a deep-seated feature in many early Mollusca.

Haemocyanin is a relatively inefficient oxygen carrying pigment, though it is far more concentrated in cephalopods than in Gastropoda. The cephalopods also have a high utilization of oxygen from the pallial water – 63% in *Octopus* as compared with 5% to 9% in sedentary lamellibranchs, though these have increased their total pallial flow for feeding purposes. The gastropods, with no such augumented gill surface as in the bivalves, and no muscular pumping as in cephalopods, pass a much smaller water volume, from which, however, they make a high oxygen extraction, estimated at 56% in a *Haliotis* species, 38% in a *Murex* and 79% in a *Tritonium*.

5 Feeding and digestion

The first molluscs probably all fed on fine particles. Food was scraped up by the broad radula, which was covered with many rows of small, uniform teeth. A good deal of sand and other indigestible matter must have been swept into the mouth along with diatoms and fragments of decaying plants. Such a mixture of particles is a far more difficult food to deal with than flesh or fluid or plant tissues. And since the early molluscs had inherited the method of intracellular digestion, there was the problem not only of collecting and transporting food, but also of grading it into coarse and fine particles, since only the smallest could be phagocytosed by the digestive gland. Continuous transport and ciliary sorting were thus among the earliest functions of the molluscan gut.

The chitons (Polyplacophora) have a strong and abrading radula, with tooth row reduced in number. Feeding is comparable with that of limpets (Docoglossa) in abrading large quantities of surface-growing nutrients. In the Aplacophora, the gut is extremely simple, with the digestive tract a straight tube, and the stomach itself lined with digestive cells. With the Caudofoveata, feeding on detritus and foraminifera in deep-water ooze, the radula has a few large pointed teeth, or even only one, directly eversible from the buccal floor. In the Ventroplicida, as in *Neomenia*, the buccal bulb is suctorial; the diet is of fluid protoplasm from hydroids and gorgonians, and the radula comprises one or several sharp teeth.

In gastropods, chitons and monoplacophorans, the complex parts of the gut are the buccal mass and the stomach. Whereas the latter is a highly specialized ciliary system, the buccal mass operates by a set of intricate muscular mechanisms.

Gastropod feeding is very diverse, and with it the radular and odontophoral structure vary greatly. The earliest radular type, among the archaeogastropods, is a wide sheet with numerous small teeth (*rhipidoglossan*), sweeping flexibly over irregular surfaces and brushing up or abrading fine particles. In the *docoglossan* radula of

D

limpets and in the *taenioglossan* (Mesogastropoda), the radula is narrower, with a smaller number of robust teeth, strongly abrading the substrate. In the *rachiglossan* radula, of Neogastropoda, the row is reduced to three sharp cutting teeth for carnivorous diet; while the teeth of the *toxoglossan* radula are highly specialized (in Conidae) for harpooning and toxin conduction with mobile prey.

The specialization of feeding habits, with the evolution of carnivorous diet, has been carried furthest in the higher prosobranchs and the opisthobranchs. The Pulmonata have remained chiefly herbivores, primitively with a broad and many-toothed radula. The early Ellobiidae and other marine pulmonates are unselective deposit-feeders, and the mud-snails (Amphibolidae) ingest fine organic deposits from the surface layers of mud. The Siphonariidae parallel the true limpets by browsing on algae, as do also the freshwater Ancylidae.

Land pulmonates have a prolific supply of plant food: leaves, shoots, berries, fruits and fungi, but above all decaying vegetable matter. The stomach is simplified and extracellular digestion is very powerful; the enzymes include cellulase, evidently the contribution of bacterial populations permanently living in the crop and intestine.

Carnivorous pulmonates have evolved several times, as with the southern Rhytididae – including its large *Paryphanta* species – feeding on other snails and slugs; British *Testacella* massively swallowing earthworms; and the large *Glandina* (southern Europe and Central America) feeding on land operculates by boring their shells in the manner of a *Natica*.

The buccal mass

The 'tongue' or odontophore lies within an arterial blood space, communicating with the haemocoele of the head. It makes two sorts of movement: backwards and forwards, of the whole structure within the buccal cavity; accompanied by protraction and retraction of the subradular membrane, covering its surface and bearing the teeth.

Graham has given a fine comparative review of the working of the buccal mass. The odontophoral protractors are a cone of muscles from the hinder ends of its paired cartilages to the anterior body wall and lips. Retractor slips run back from the posterior ends of the cartilages to the body wall. The sub-radular membrane has lateral and median protractors and radular tensors, coordinating the movements of both membrane and radular sac. The teeth are erected, as they move over a bending plane, by the diverging of the dorsal edges of the cartilages by ventral approximator muscles. Radular retractor muscles form massive straps running between the cartilages to the membrane and radular sac.

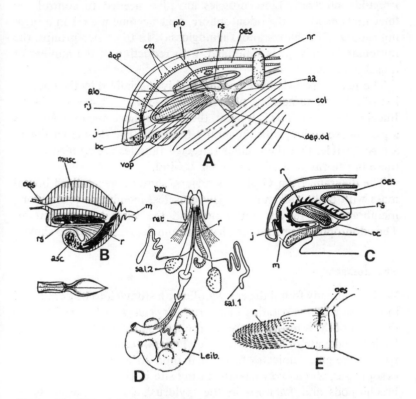

Figure 32 The gastropod buccal mass
(A) Schematic prosobranch buccal mass (*based on Graham*)
(B) Sacoglossan buccal mass, of *Alderia*; (*inset*) a single tooth
(C) Generalized longitudinal section showing odontophore and jaw
(D, E) Buccal mass of the predaceous *Testacella*, with external musculature and odontophore extruded
aa, anterior aorta; alo, anterior levator of odontophore; asc, ascus sac; b.c, buccal constrictors; cm, circular muscles; col, columellar muscle; dep.od, depressor muscles of odontophore; dop, dorsal odontophoral protractors; j, jaw; m, mouth; musc, superficial musculature; nr, nerve ring; oc, odontophoral cartilage; oes, oesophagus; plo, posterior levator of odontophore; r, radula; rj, retractor of jaw; rs, radular sac; vop, ventral protractors of odontophore

In addition to these, numerous ancillary muscles act as tensors, especially in rhipidoglossans, and in chitons and monoplacophorans, which lack the flexibility given by a mobile head. In rhipidoglossans, where the radula is a brushing organ making close contact with an irregular substrate, these muscles are also needed to control the finer movements of the odontophore, that become useless in a rasping radula (Docoglossa and Taenioglossa). In these two groups, the muscular assemblage is simplified, with reduction of the number of muscles and cartilages.

The neogastropods have become highly specialized by the possession of a long and narrow, pleurembolic proboscis. Especially in the Buccinacea, where this is longest, the odontophore narrows, and has a piston-like action, with the mouth terminal. The slender cartilages are so flexible, that the radular retractor muscles are applied not to these but to the wall of radular sac behind.

The diagram, from Graham's review, shows a generalized buccal mass with body wall muscles, jaw muscles, membrane protractors, membrane and radular retractors and the ventral approximator. The odontophoral protractors, levators and depressor are also shown.

The stomach

We have already found the early molluscan stomach to be dominated in its action by the rotating faecal rod, or *protostyle,* lying within the first part of the intestine known as the *style sac.* Many microphagous feeders make use of some kind of a windlass, turned by the cilia of a part of the gut. Tunicates form their food string by twisting in the oesophagus, *Amphioxus* has a rotating site in the colon, polyzoans, brachiopods and *Phoronis* in the 'pylorus'. Only in the molluscs, however, do the style and the ciliated walls of the stomach take on such specialized functions, and only here – both in non-carnivorous mesogastropods and in higher bivalves – does a *crystalline style* develop, carrying in its substance a digestive amylase.

In primitive bivalves (Protobranchia) there is a short, stout faecal protostyle, found by Owen already to contain amylase. *Neopilina* has also a protostyle, though Polyplacophora and also limpets (Docoglossa) have lost this early mechanism.

Whether faecal or 'crystalline' (actually of flexible, hyaline globuloprotein) the style head projects into the stomach, and a food string from the oesophagus is in many species wound on to it, forming a tight spiral and in this way drawn short and pulled gradually

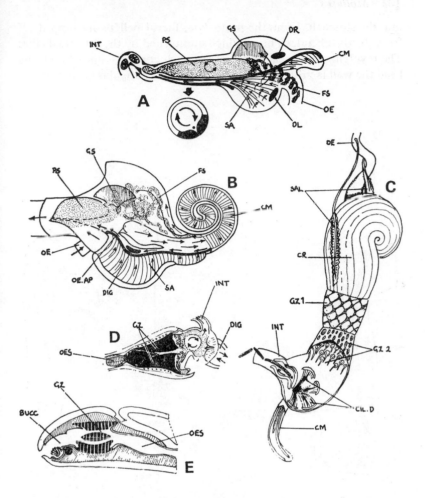

Figure 33 The gastropod stomach and foregut
(A) Idealized stomach of early prosobranch
(B) *Turbo*, with spiral gastric caecum
(C) *Aplysia*
(D) *Limacina*
(E) *Philine*
LS, buccal mass and gizzard; CIL, ciliated tracts leading to digestive diverticula; BUCC, buccal mass; CM, caecum; CR, crop; FS, food string; GS, gastric shield; DIG, digestive diverticulum; DL, DR, left and right digestive diverticula; GZ, gizzard; GZ1, GZ2, first and second gizzards of *Aplysia*; INT, intestine; OE.AP, oesophageal aperture; OES, oesophagus; PS, protostyle; SA, sorting area; SAL, salivary glands

into the stomach. From the protostyle, faecal pellets are nipped off from the distal end, to be moulded and turned by the intestinal cilia. The revolving style head stirs the stomach contents, and close to its head the wall is protected from abrasion by a cuticular gastric shield.

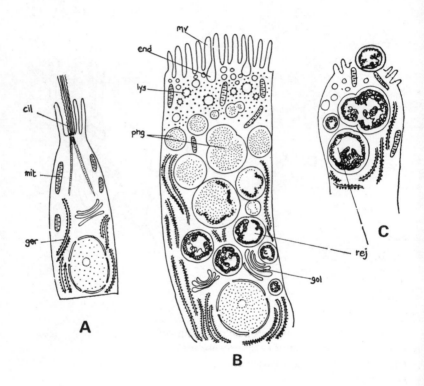

Figure 34   The molluscan digestive cell
(A) Young cell with cilium
(B) Mature absorbing stage
(C) Post-mature extruding phase
end, endocytotic vesicle; cil, cilium; ger, granular endoplasmic reticulum; gol, golgi apparatus; lys, absorption vesicle with attached lysosomes; mit, mitochondria; mv; microvilli phagosomes; rej, rejection vesicles

The rest of the stomach is lined with fine ridges and furrows, acting as a ciliary sorting area, of the kind already seen on the labial palps. As the food string rotates near this area, particles are all the while detached. At least in the more elaborately developed sorting areas, coarse material is flung into the deeper grooves and taken by cilia to the intestine. Finer particles kept in suspension by cilia on the ridges gradually reach the paired digestive diverticula. With the lower pH of the stomach fluid the mucus of the food string becomes less viscid (5 to 6), which assists the shedding of its particle load. The intestinal pH is higher, about 8, and the viscous mucus here binds the contents into firm pellets.

The digestive gland

The two digestive diverticula lead from the stomach to much-branched glandular follicles. These exhibit a cycle of absorption and secretion. Most of the digestion being intracellular, the absorbing cells produce a protease, amylase and lipase. Extracellular enzymes are available too, secreted by the oesophagus in glandular pouches opening singly or in series into its middle length. The salivary glands may produce enzymes as well, particularly protease in carnivores. In both gastropods and lamellibranchs that possess a crystalline style in the style sac, this rod liberates its own store of carbohydrate-splitting enzymes as its substance dissolves in the stomach. Recent work suggests that an amylolytic enzyme may sometimes exist along with a protostyle. Finally, the amoebocytes of the blood stream play an important part in digestion. They migrate freely into the gut from the blood system and phagocytose particles within the stomach. They may then retreat back into the epithelium, or surrender their contents for final absorption by the digestive gland.

The fine structure and function of the digestive gland cells have been given much recent study. The absorbing cells have been found to undergo a sequence of stages from (i) undifferentiated 'crypt cells' bearing a few long cilia, giving rise to (ii) actively absorbing cells with a micro-villose border. Here, endocytotic vesicles can be seen to form, with fine particulate food carried into the cell. Primary lysosomes, produced from a Golgi system, attach to these, and pour their enzyme contribution into what now become large phagolysosomes. Further digestion follows, with absorption of the products, after which the remains are held in 'residual vesicles', ultimately discharged into the lumen by the fragmenting of the tips of the absorbing cells (stage (iii)). As well as undigested residues, it would appear that enzymes are also liberated into the stomach, to become active in the next round of preliminary extra-cellular digestion. Their waste material is carried to the intestine, or impacted upon the protostyle. After discharge, stage (iii) cells are replaced by smaller ones from the crypts.

Later Prosobranchia

Only slight changes are needed to convert a microphagous gastropod into a macroherbivore, a ciliary feeder, or even a carnivore. The early Mesogastropoda were destined to a wide radiation in feeding habits, and their evolution is in great part a story of changing diets.

Prosobranchs have become carnivores in several ways, and the gut usually shows a striking contrast with that of herbivores or ciliary feeders. The diet is smaller in bulk and feeding is intermittent. The style sac and sorting area disappear and the stomach itself is reduced to a simple bag into which enzymes pass from the digestive gland. Digestion is wholly extracellular, the digestive gland merely secreting enzymes and absorbing. It is now the mouth parts and the buccal mass which become specialized for the harder task of procuring living animal food. The radula, while short, is equipped with sharp, cutting teeth. In the carnivorous Neogastropoda, each row is reduced to three strong teeth, or sometimes only one (Fig. 35G). The salivary and oesophageal glands enlarge for the secretion of protease, and become stripped away from the gut wall, connected with it by long narrow ducts. In this way the whole anterior gut is free to slide forward through the nerve ring, as the proboscis – with the buccal mass at its tip – is everted like the finger of a glove. In some families the extruded proboscis may reach four or five times the length of the rest of the animal.

Many families have developed carnivorous ways by easy stages, grazing not on algae but on sponges and other encrusting animals. *Diodora* – among the Archaeogastropoda – already shows this trend. The Cypraeacea or cowries – primitive Mesogastropoda – are another good example. Tropical cowries are generally grazing herbivores with style sac, sorting area and gastric caecum, living on cropped algae or bottom deposits. The small British *Simnia* has come to feed on the tissues of the coelenterates *Eunicella* and *Alcyonium*, crawling over them and grazing with a short snout. *Trivia* has next developed a proboscis, and feeds at leisure on the compound ascidian

Diplosoma, eating the zooids as it ploughs them out of the test. Finally, *Erato* dips its long proboscis expertly into the mouths of zooids of *Botryllus* and *Botrylloides*, cleaning out the tissues from within. The Lamellariidae are a related family feeding on ascidians; *Cerithiopsis* grazes on sponges and *Bittium* selects foraminifera from the bottom deposits.[131]

Such stationary prey needs no active pursuit and one family, the Ianthinidae, has carried the habit of feeding on coelenterates from benthic to pelagic life. These violet snails live by exploiting the drifting siphonophore, *Porpita*, and *Velella* clinging beneath the disc for transport when not using their own raft, and feeding on it as well, rasping away the tissues to the bare chitinous 'skeleton' of the siphonophore.

A sluggish carnivorous life points the way to parasitism. First, some sedentary gastropods have become permanent associates of other animals. *Hipponyx* has the harmless habit of settling on the shells of *Turbo* and eating its faecal pellets. That such a habit is an ancient one is shown by the Palaeozoic gastropod *Platyceras*,[8] which had already settled on the calyx or arms of crinoids often near the anus and evidently lived too on faeces. Modern capulids carry this habit further, attaching near the edge of a mussel or scallop shell. *Capulus ungaricus* is a ciliary feeder, intercepting particles from the ingoing feeding current of the host. It also has a suctorial tendency; and leads on to the limpet-like *Thyca* which plunges its proboscis into the tissues of starfish and echinoids. Such forms perhaps initiated the long series of endoparasitic mesogastropods living in echinoderms (see p. 196).

Ectoparasitic gastropods are of more or less normal appearance. There are two families to be considered, feeding on the blood and body fluids of sedentary invertebrates, such as worms, echinoderms, bivalves and crustaceans. In the Eulimidae, associated with echinoderms, the proboscis is short. Jaws and radula are both lost, and the pharynx forms a muscular pump with proboscis glands by which the tissues of the host are softened. The Pyramidellidae – which Fretter and Graham have suspected to be opisthobranchs – have elaborate mouth parts, exquisitely modified for piercing and sucking. The chitinous edges of the jaws are opposed to each other and prolonged into hollow stylets, running along a tube formed from the buccal lips. The stylets are divided so as to form an upper suctorial and a lower salivary canal. The radula is lost, all hard parts being formed from the jaws. The muscles of the pharynx form a sucking pump, and the ampullae of the salivary glands are muscular as well, providing a pump for the enzyme-carrying saliva. With a diet of fluid protein the rest of the gut is exceedingly simple.[137]

Not all mesogastropod carnivores have sluggish habits. The fast-swimming Heteropoda, with movable telescopic eyes, are active predators, catching medusae, small fish, copepods and pteropods and themselves falling prey to larger heteropods. The large *Ptero-trachea* and *Carinaria* carry the buccal bulb at the end of their flexible 'trunk', periodically thrusting out the sharp-toothed radula to seize prey. They will even attack the human finger-tip.

The Neogastropoda have carried flesh-eating to a high level; and with them we must mention some remaining mesogastropods. There are three tribes of whelk-like Neogastropoda, the Buccinacea, Muricacea and Volutacea (together forming the Rachiglossa), and a fourth tribe, the Toxoglossa, composed of very specialized carnivores, represented by the Conidae and Terebridae and in Britain by the Turridae.

A typical rachiglossan radula, with three sharp-cusped teeth, is shown in Fig 35G. The Buccinacea are the least specialized of Rachiglossa; they feed on dead or decaying animal matter, as in the Nassariidae, or frequently, as in the Buccinidae, on living flesh. The anterior shell canal and the osphradium are always well developed, and are used in detecting food. With some species of *Alectrion*, if a piece of decaying meat is placed between glass sheets the snails converge from a good distance around this 'sandwich' and insert the narrow proboscis to a length of 12 cm between the sheets, till the radula at its tip is able to rasp the food.

Such a use of the proboscis foreshadows the habits of the Muricacea, many of which are shell-borers, feeding on live bivalves and gastropods. They drill a neatly chiselled hole through which their long proboscis is inserted to reach the tissues within. In muricaceans such as *Urosalpinx*, *Ocenebra* and *Nucella* boring is chiefly mechanical. The shell is gripped with the foot, and the odontophore extruded. Short strokes of the radula abrade the shell, a few strokes in one direction and then – after twisting through an angle – a few in another, the hole being gradually deepened. These genera have an accessory pair of salivary glands absent in the whelks, and opening on the edge of the mouth. They may well play some part in boring, perhaps assisting by their secretion in softening the calcium carbonate during excavation. *Nucella lapillus* feeds on both mussels and acorn barnacles. The latter are not drilled but are smothered with the foot, till the valves can be dislodged. The same method is used

by some species of *Thais* with gastropods such as trochids and tur-
binids, which are gripped with the foot until the columellar muscle
relaxes and the operculum can be dislodged. Some species of thaids
have a strong tooth at the rim of the shell, which they are credited
with using as an oyster knife. *Murex fortispina* in New Caledonia

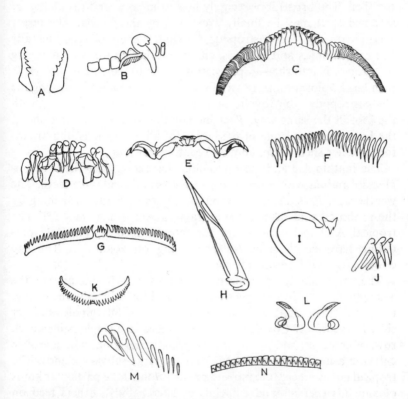

Figure 35   Representative example of the radula in Amphineura and
Gastropoda

(A) *Lepidomenia* (Aplacophora)
(B) *Boreochiton* (Polyplacophora)
(C) *Margarita* (Rhipidoglossa)
(D) *Patella* (Docoglossa)
(E) *Vermetus* (Taenioglossa)
(F) *Ianthina* (Ptenoglossa)
(G) *Fasciolaria* (Rachiglossa)
(H) *Conus* (Toxoglossa)
(I) *Scaphander* (Cephalaspidea)
(J) *Doris* (Doridacea)
(K) *Aeolidia* (Aeolidiacea)
(L) *Goniodoris* (suctorial
       Doridacea)
(M) *Rhytida* (carnivorous
       Stylommatophora)
(N) *Helix* (herbivorous
       Stylommatophora)

takes *Arca* in the same way. The Coralliophilidae are stationary pur-purids which have lost the radula and feed suctorially on coral tissues. From them has arisen the uncoiled *Magilus,* evidently a ciliary feeder. One group of mesogastropods, the Naticidae, prey upon burrow-ing bivalves, which they bore by the acid secretion of a small gland on the underside of the tip of the proboscis as it is pressed against the shell. The broad foot securely holds the prey, and the etched or softened shell may be finally excavated by the radula. The larger sand-burrowing Mesogastropoda, Cassididae and Doliidae, generally live on echinoids and bivalves, either smothering them or plunging the proboscis into the soft parts, as does the tun-shell *Dolium perdix* with large holothurians. Most of the tropical Volutacea – among the Neogastropoda – burrow in sand and take prey by smothering with the foot in the same way. They include the volutes, the olive shells, the harp shells and one of the largest of all gastropods, the broad-footed baler shell *Melo,* about 45 centimetres long.

The Ianthinidae and the Epitoniidae have a characteristic 'pteno-glossan' radula with a broad expanse of small blade-like teeth. The wentletraps, *Epitonium,* live upon anemones or corals, attaching by the proboscis and rasping tissue fragments in similar mode.[274] The tropical Architectonicidae are coelenterate feeders, *Philippia* pre-dating hard corals, and *Heliacus* living on zoanthids. The egg cowries (*Ovula*) feed on soft corals.

The most highly specialized carnivores are found among the Toxoglossa or 'arrow-tongues'. Central and lateral teeth are lacking, the radula (Fig. 35H) consisting of long slender marginals at either side. In the family Conidae these are reduced to a single pair in each row, of great size and prolonged into slender harpoons with a groove carrying a neurotoxic secretion from the large salivary glands. The tropical cones all capture moving prey.[185] Some take particular kinds of annelids (nereids and euniciids and terebellids), others feed on bulloid opisthobranchs. *Conus marmoreus* feeds on other cones, and four species – *C. cattus, C. striatus, C. tulipa* and *C. geographus* – catch live blennies and gobies. The prey is located by the testing of the pallial current by the large osphradium, after which the cone crawls towards it and 'covers' it with the poised proboscis. During a successful strike a single tooth is everted which harpoons the fish and injects saliva. Its struggles stilled, the whole fish is quickly en-gulfed by the distended proboscis. Digestion is rapid, beginning while most of the fish is still in the pharynx and crop. If the first strike fails, a second tooth is employed for another attempt. Most Toxoglossa appear to produce poisonous saliva; and that of the Australian *C. geographus* has proved fatal to man.

Opisthobranchia

Many of the feeding habits shown by the Prosobranchia have been paralleled or improved upon by the opisthobranchs, and these gastropods are above all masters of the grazing carnivorous and suctorial life. As in prosobranchs, and in pulmonates too, the earliest opisthobranchs are microphagous browsers. Primitive bulloids such as *Actaeon* have a broad radula with small uniform teeth, and traces of the style sac and sorting area. At an early stage, however, both opisthobranchs and pulmonates cease to depend much on mucus and cilia, and the gut becomes more muscular.

Most of the Bullomorpha have a large oesophageal gizzard, furnished with chitinous or calcified tooth plates, while the stomach is small and simplified, little more than a vestibule for the digestive diverticula.[127] Just as in prosobranchs, browsing on detritus has led to carnivorous ways, especially in those bulloids that can invest their prey with the broad foot. While *Actaeon* and many species of *Haminea* are deposit feeders, most forms have acquired strong sickle-shaped radular teeth (Fig. 35 I) for seizing shelled prey, which is then crushed in the gizzard. Thus – in a series of descending size – *Scaphander* swallows whole bivalves and gastropods, *Philine* feeds on *Nucula* and young bivalves, *Retusa* on *Hydrobia ulvae* and *Cylichna* on foraminifera.

A special offshoot of the bulloids is seen in the thecosomatous pteropoda, which are small pelagic ciliary feeders. The gill is lost, but the pallial mucous gland and ciliary currents on the wings and mantle collect food. The bulloid gizzard is still present, an unusual organ – it was supposed – in a ciliary feeder, until it was found in *Limacina* to form a tiny mill for crushing the cases of diatoms.[219]

The Anaspidea (or Aplysiomorpha) are almost the only macroherbivores among the opisthobranchs.[168] The majority feed by cropping living seaweeds with their paired jaws and broad radula. The gut is more complicated than in bulloids, with a storage crop and the gizzard divided into two chambers – an anterior one for masticating and a posterior one with delicate teeth for straining. The stomach is reduced: its posterior caecum – when present in opisthobranchs – serves not for sorting but for fashioning faeces (Fig. 33).

The gymnosomatous pteropods are a law unto themselves. They

are sometimes held to have arisen from aplysioids, but in feeding bear no resemblance to them or to any other opisthobranchs. They are active predators like heteropods, living chiefly off the schools of

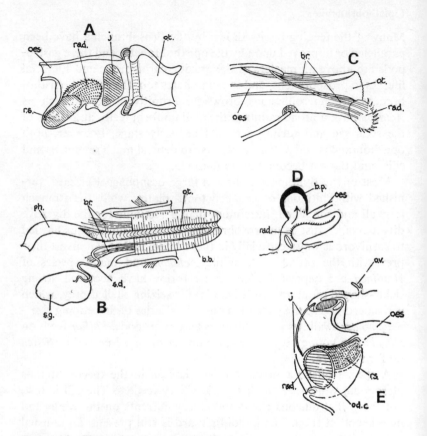

Figure 36   The opisthobranch buccal mass
Sagittal sections from (A) *Chromodoris*
(B) *Dendrodoris*
(C) *Gymnodoris*
(D) *Okenia*
(E) *Tritonia*
b.b, buccal bulb; b.p, buccal pump; b.r, buccal retractor muscles; j, jaw; od.c, odontophoral cartilage; oes, oesophagus; o.t, oral tube; o.v, oral veil; ph, pharynx; rad, radula; r.s, radular sac; s.d, salivary duct; s.g, salivary gland

thecosome pteropods with which they are always found. *Clione* and *Pneumodermopsis* are familiar Atlantic examples. All their weapons are concentrated in the buccal mass, which forms an elaborate armoury indeed. The radula has sharply pointed exsertile teeth. It is reinforced by a dorsal chitinous jaw, and by a sheaf of prehensile hooks, carried in an eversible pocket at either side of the buccal cavity. In addition, the Pneumodermatidae have a set of branched buccal tentacles, clustered with stalked or sessile suckers, while the Clionidae show a circlet of adhesive oral papillae, the *cephaloconi*.[224]

Elsewhere the opisthobranchs can show few active predators. Of those that catch moving food the most interesting are the slow-swimming nudibranchs of the Tethyidae (*Melibe* and *Tethys*). These have a wide cowl over the head which is thrown from side to side as they swim, to collect amphipods, isopods and large copepods. The gizzard of *Melibe* is often crammed with crustacea up to 2·5 cm long, while the large species of *Tethys* feed on *Squilla*.

Dorid feeding

The Doridacea, the largest group of the nudibranchs, have a wide radiation of feeding habits, best reviewed by D. K. Young[343] from his study of forty-eight Indo-West Pacific species. Five basic groups will be recognized here.

(i) *Rasping sponge-feeders* including Dorididae (nine sub-families) and Hexabranchidae. These are the most generalized, with a strong rasping radula (comparable in action with the Taenioglossa), odontophoral cartilages and (sometimes) jaws. The stomach is a large sac, vigorously churning the food for extracellular digestion. Fluid food is then absorbed by the digestive gland and its breakdown completed intracellularly. A stomach caecum compacts residual sponge spicules into faeces.

(ii) *Sucking sponge-feeders* (Dendrodorididae). The odontophore and radula are lost, replaced with a suctorial oral tube. Instead of radular action, there is a chemical attack on food, followed by mechanical sucking. A large 'ptyaline gland' lies under the buccal bulb, with a median duct opening ventrally into the mouth. Its secretion is slightly acid, but free of enzymes, which are produced only by the digestive gland.

(iii) *Suctorial feeders on polyzoans and ascidians*. This group – not dealt with by Young – contains the Goniodorididae. The European *Acanthodoris pilosa* and *Onchidoris* feed on *Alcyonidium*, and

Goniodoris species on *Botryllus* and *Dendrodoa*. Each radula row has only four teeth, two of them curved and sharply serrated (Fig. 34L). Only fluid food is ingested, sucked up by a powerful buccal pump on the roof of the pharynx. Its movements are co-ordinated with the opening and closing of the mouth in pumping food down the oesophagus. The buccal pump shows every stage (Fig. 36D) from *Acanthodoris* with a shallow open chamber, to *Onchidoris fusca* where it is like a small pea attached by a narrow duct. The rest of the gut is simplified, as in Dendrodorididae, with the faeces of small bulk, merely a yellow outflow from the digestive gland.

(iv) *Engulfing opisthobranch feeders* (Polyceratidae). Small *Polycera* species engulf and chew polyzoans, such as *Bugula*. Of the tropical *Gymnodoris* species, Young has shown *G. okinawae* to feed on Elysiidae and other sacoglossans, and *G. bicolor* on other *Gymnodoris*. A short oral tube leads to a dilated buccal vestibule, beyond which is the buccal mass. The radula is short and broad, and its teeth longer and sharper than in dorids. This group has marked analogies with other engulfing gastropods: the prosobranchs *Ianthina*, the bulloid *Philine* and the carnivorous pulmonate *Testacella*.

(v) *Boring polychaete feeders*, for example the small red nudibranchs *Okadaia* (family Vayssiereidae). These feed on spirorbid tube worms, first boring a neat hole in the posterior part of the tube, by the serrated first lateral teeth, then inserting the proboscis for engulfing, with the elongated second lateral teeth serving to grasp the prey.

Pleurobranchoids such as *Oscanius* and *Pleurobranchaea* insert the long suctorial proboscis into a hole rasped in an ascidian test by the numerous radula teeth. A dorid ascidian prober, *Dendrodoris citrina*, has entirely lost the radula and digests a proboscis entry into the test.

The tritoniids feed on sessile coelenterates and this habit has been passed on to aeoliids, which have acquired a remarkable adaptation of the gut.[153] The digestive diverticula branch into a series of tubules lying in the club-shaped cerata on the dorsal surface of the body. Opening by a pore at the tip of each ceras is a small cnidus sac, separated by a sphincter from the digestive gland. In the epithelium lie numerous undischarged nematocysts, half a dozen or more in each cell, where they have been stored from the coelenterate prey. Such an association of aeoliids and nematocysts poses many questions. How does the mollusc remain immune to them? Discharge is probably prevented by the low pH of the gut and the chitinous coat of the nematocyst, wrapped in mucus. They come into action only when the animal is injured and a ceras forcibly detached. Though they cannot be discharged spontaneously, the nematocysts must still be of great use in giving immunity from predators. With them – as we have seen – the aeoliids have developed pronounced warning colours of extreme beauty.

The last group of opisthobranchs, the Sacoglossa, are also suctorial, but in a very different way from any other gastropods. They are herbivores living on the cell-sap of green algae, each species generally specific to one kind. *Hermaea, Caliphylla* and *Elysia* take *Codium* and *Bryopsis* species, *Actaeonia* feeds on *Cladophora*, and at their tropical zenith sacoglossan species abound on the forms of *Caulerpa*. The large algal cells are lanced one by one, as the thallus or filament passes between the lips. The radula (Fig. 32B) has a single tooth in each row, a small blade fashioned precisely to the cell size involved. The radula lies in a peculiar ⊃-shaped tube, with back-directed limbs and opening at its bend into the pharynx floor. Only one tooth is in use at a time, and these originate in a continuous series in the upper limb, the worn teeth being stored when detached in the lower limb, known as the 'ascus sac'. The pharynx constitutes a powerful force pump. Its roof is highly contractile and may lead into muscular pockets. These work in concert with an oesophageal valve into the stomach. The food is entirely fluid and the stomach forms a mere vestibule for a spacious digestive gland, in some species with cerata.

Much recent study has been given to the symbiosis of Sacoglossa with the chloroplasts ingested from the food plant. Sucked in undamaged, these persist in the cells of the digestive diverticula. In British *Elysia*, Taylor in 1970 showed they become organelles.[287] In Pacific *Placobranchus* and *Tridachiella*, Trench and Muscatine found they had the same photosynthetic capacity as intact zooxanthellae, taking up labelled $^{14}CO_2$ and producing labelled glucose. Their action may be compared with the symbiotic nutrition of giant clams (p. 118).

Scaphopoda

The tusk shells are detritus feeders and microcarnivores. The captacula arise in dense clusters at either side of the head and stream out widely through the sand to search for organic particles often including foraminifera and sometimes bivalve spat. These are carried down the tentacle by ciliated tracts.[105] The pedal lobes and shorter captacula are used in grasping large collected particles and bringing them to the mouth. In *Dentalium entalis* up to a hundred forams may be crammed at one time into the cavity of the broad, flat proboscis. The radula is of relatively immense size with five teeth in each row. Strong erectile laterals seize each foram in turn and remove it to the buccal mass. The oesophageal pouches are small and apparently do not secrete; there are no salivary glands. The stomach is a muscular bag where food is triturated and undergoes extracellular digestion. There is no style sac and only vestiges of caecum, sorting area and shield.[226]

Bivalvia

Lamellibranchs are almost all ciliary feeders on suspensions or deposits. Food is collected by the gill and labial palps, and no bivalve has a radula, buccal mass or salivary glands. The oesophagus is short and the intestine a mere transport tube for faeces. The stomach and style sac are, however, much more complex in lamellibranchs than in any other molluscs (Fig. 37). The evolution of the sorting areas and sorting caecum, and the crystalline style with its enzymes, shows a striking parallel with what we have seen in ciliary-feeding gastropods. In both groups the functions of the stomach are to sort a mixture of particles, to carry what is worthless to the intestine, to convey fine fragments after preliminary extracellular digestion to the digestive diverticula, and to ensure that waste matter returning from the diverticula reaches the intestine without remixing with the food. It is a little difficult to explain the intense specialization of the stomach wholly in terms of the needs of a ciliary feeder. By comparison with deposit-feeding gastropods the food is small in bulk and already partly sorted on the gills and palps. One might have thought the bivalve stomach had a simpler task to perform. Its evolution may, however, be only one part of a general trend of advance repeatedly found in this highly specialized group. Its pattern is beautifully adapted to functional needs, perhaps not wholly conditioned by them.

The researches of Yonge[316] and Graham[155] and later of Owen[247] and Purchon,[263] have taught us much about the working of the stomach in lamellibranchs. The whole action is dominated by the rotation of the crystalline style, which – except in the protobranchs and septibranchs – is a long, flexible rod, hyaline and usually colourless. As in style-bearing gastropods (p. 100) it is built up of concentric layers of muco-protein, secreted by the edge of the typhlosoles that cut off the style sac from the intestine. In many higher families ('eulamellibranchs') the style sac is cut off completely from the intestine by the fusion of its bounding typhlosoles: in Anisomyaria such as Pectinidae, Ostreidae and Mytilidae, the sac communicates by a slit with the intestine, and the style may become permeated with minute food particles. These have been 'retrieved' by catching up in the viscid style substance as they travelled along the intestine, and are thus returned to the stomach for further digestion. The crystalline style always contains amylase and glycogenase, which are set free in the stomach as part of its head dissolves; in some species a cellulose-splitting enzyme has recently been identified as well. The head of the rotating style bears upon or is partly surrounded by a flange of protective cuticle of the stomach wall, known as the gastric shield. After removal from water or cessation of feeding, especially where the style sac is open to the intestine, the whole style gradually dissolves, and some high tidal bivalves – such as *Lasaea rubra* – pass through a regular cycle, in which the style almost disappears when the tide is out and is later resecreted.[223]

As well as being an enzyme store, the style keeps its original function of a stirring rod and a windlass. One or more food strings from the oesophagus are wound on to it, or thrown by it into a tight spiral. As these are drawn into the stomach, particles from them are shed, partly by contact with the stomach wall but mainly by effect of the lowered pH of the stomach contents in reducing the viscosity of the mucus. Moreover the free particles in the stomach are repeatedly swept across the ciliary sorting area. The lightest are kept in suspension by cilia at the crests of the ridges and undergo preliminary extracellular digestion, both by the style enzymes and the amoebocytes. These are finally conveyed, minutely divided, to the digestive diverticula where they are phagocytosed and digestion is completed intracellularly.

As in early gastropods, most bivalves show a gastric caecum, which may be a feature of the earliest molluscs. Into this pouch a portion of the sorting area and the often numerous openings of the digestive diverticula retreat, together with a prolongation of one of the typhlosoles bounding the style sac from the intestine. Within the caecum this typhlosole frequently forms a tube within a tube (Fig. 37B) sending a branch into each diverticular aperture. The inner tube is the incurrent passage to the diverticulum, into which sorted food travels for absorption. The outer tube is an excurrent passage carrying waste material returned from the digestive cells to the intestinal groove. Along the ducts of the digestive gland, the outgoing

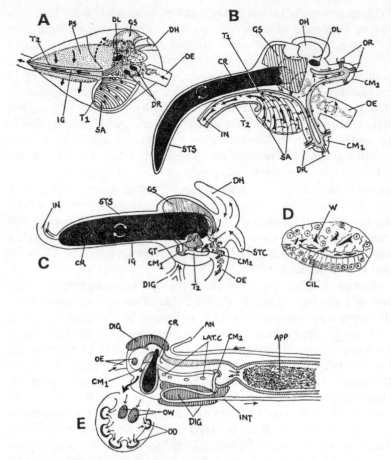

Figure 37 The bivalve digestive system
Stomachs of (A) *Nucula* with short faecal protostyle
(B) A suspension-feeding eulamellibranch
(C) Deposit-feeding *Tellina*
(D) Digestive tubule of *Teredo*
(E) Stomach of *Teredo*, with saccular appendix containing wood fragments
APP, appendix storing wood; AN, anus; CIL, ciliated cells; CM 1,2,
caeca of the stomach; CR, crystalline style; DH, dorsal hood; DIG,
digestive diverticula; DL and DR, left and right digestive diverticula;
GS, gastric shield; GT, gastric tooth; IG, intestinal groove; IN, intestine;
OE, oesophagus; PS, protostyle; SA, sorting area; LAT. C, lateral caeca;
STC, storage caecum; STS, style sac; T1, T2, typholosoles 1 and 2; W,
wood fragment phagocytosed; OD, opening to normal digestive cells;
OW, opening to wood absorbing cells

passage alone is ciliated, the ingoing current proceeding by counter-flow to the passage of particles outward (see Owen).[247] The terminal tubules of the digestive gland (c.f. Fig. 34) contain young, non-absorbing cells, often bearing long cilia, and mature cells which phagocytose particles from the lumen and become filled with greenish or yellow food vacuoles. Residual waste is extruded by the fragmentation of the tips of the mature cells to form spheres packed with coloured vacuoles which are returned to the stomach. Traces of enzymes may by the same means be made available for extracellular digestion. The intestinal groove is usually, however, closed from the stomach by a valve formed by the typhlosole, so that rejecta from the sorting areas is joined by waste from the digestive gland, without contact with food in the stomach.

Graham, Purchon and Owen have shown in detail how the stomach may become modified in higher eulamellibranchs. Here the sorting caecum is frequently divided into two, both opening separately into the stomach. They receive the multiple apertures of the digestive diverticula and a prolongation of the typhlosole now winds its path through both caeca. In certain deposit-feeding groups of lamellibranchs the stomach may serve as a triturating region, particularly in the Tellinacea which combine a very massive crystalline style with a heavy gastric shield.

The wood-boring *Teredo* has a capacity to digest cellulose, and the crystalline style is notably small.

Teredo navalis has been shown to have a daily rhythm of alternating sorts of activity. Siphoning of a water current occurs during the day, the pallial systems being quiescent at night, when tunnelling takes place. Wood is ingested at night, but plankton is collected from the inhalant water during the day. *Teredo* is thus a continuous feeder, with spells of nourishment derived from two sources. Little extracellular digestion goes on in the stomach, which is almost exclusively a storage organ, for large amounts of unsorted material from the sea or the tunnel. A very large stomach appendix serves as a temporary store for large wood fragments, that are not – as would normally happen – rejected by the labial palps. The digestive diverticula have developed special areas for the phagocytosis and intracellular digestion of wood fragments. B. S. Morton has recently described their ultrastructure.[213]

Tridacnid clams have vast numbers of zooxanthellae (*Gymnodynium microarticulatum*) in all their light-exposed tissues. Division stages abound. The algae are common too in the visceral mass, contained and digested within amoebocytes, though they are never found in the digestive cells in the gut.

Zooxanthellae have always been considered a major source of nutriment, allowing the clams to attain their giant size. Recent studies by the Goreaux[148] have been directed to the utilization by the Red Sea *Tridacna maxima* of their products of photosynthesis. Specimens were exposed to labelled $^{14}CO_3$ and tritiated-DL-leucine to determine the significance in nutrition of intact and photosynthesizing zooxanthellae and of dissolved amino acids. After 10 minutes, the zooxanthellae in the siphon lips were strongly labelled, and this effect later goes deeper, to those in blood sinuses and viscera. For 48 hours after exposure, the zooxanthellae continued to release labelled substances. Strong radioactivity was found in the glandular areas, the periostracal groove, and very enlarged pallial mucous gland and ctenidial glands, and in the minor typhlosole and byssal gland.

In line with recent work on symbiotic chloroplasts in the sacoglossan slug *Tridachia*, the tridacnids seem to gain greatest benefit from photosynthates rather than by digesting older cells in amoebocytes, which may, however, provide some energy.

Labelled leucine can enter the body by exposed microvillose surfaces in mantle cavity and gut. Giant clams thus have a capacity to assimilate dissolved organic matter, but in this have no apparent superiority to any other bivalves.

In many bivalves, the rhythm and periodicity of valve adduction is now known to have a profound effect on feeding, especially the rhythmicities transmitted from the cycle of the tides. In a number of intertidal bivalves – for example *Lasaea rubra* and *Cardium edule* – the style is resorbed at low tide and resecreted at high. The histology of the digestive gland has been amplified by much electron micrograph study, and its cycle of action shows a clear sequence of phases: (i) cell formation and proliferation, (ii) active absorption and phagocytosis, (iii) digestion in phagolysosomes, and assimilation (iv) cell breakdown and extrusion of fragments.

Important reviews of bivalve digestion by both Purchon and B. S. Morton have helped to clarify the action sequence. It is clear that feeding and digestion together form a dynamic process, both spatially and temporally (see Fig. 38). Food arrives in the stomach when the style is wholly or partially dissolved (A). Fragments with enzymes extruded from the digestive gland and enzymes released from the style help break this food down. Partly digested food is then sorted in the stomach, and large unwanted particles (UF) are then

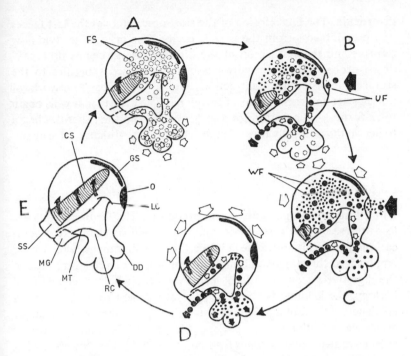

Figure 38 Cycle of functions of the bivalve stomach and digestive diverticula (*from B. S. Morton after Owen*)
Stages A to E explained in text
CS, crystalline style; DD, digestive diverticula; FS, fragmentation spherules from digestive gland; GS, gastric shield; LC, left caecum; MG, mid-gut; MT, major typhlosoles; RC, right caecum; SS, style sac; UF, large unwanted particles; WF, small food particles

passed to the mid-gut (MG) in the intestinal groove of the major typhlosole (MT) (B,C,D). Food material of acceptable size is passed to the digestive diverticula for further extracellular digestion, absorption and phagocytosis, intracellular digestion and final assimilation. Passage of food to the diverticula may be assisted by pressure from phasic contractions of the adductor occurring at this time. (D) When the bivalve ceases to feed, the mouth (O) shuts, the remaining waste is passed to the mid-gut, and remaining food to the diverticula. The final closing of the shell now removes the last faeces and pseudofaeces from the mantle cavity. The crystalline style now reforms and the epithelium of the digestive gland begins its break-down process (E), eventually passing fragmented spherules to the stomach, assisted by contraction of the 'string-bag' network of muscles around each acinus of the gland. The diverticular cells begin to re-form in preparation for another cycle, and the spherules begin to act on the now fully formed style, causing it to dissolve once again.

Cephalopoda

Almost all cephalopods are active predators. The feeding habits are less diverse than in gastropods, and the gut shows elaborate modifications for carnivorous life. Unlike the lamellibranchs, which have emphasized the stomach alone, the cephalopods show every part of the digestive system well developed. As in the whole of cephalopod biology, the keynote is speed and the attainment of a new level of efficiency. The familiar squids are pelagic, feeding on fish, larger crustacea and other cephalopods; cuttlefish take fish and crustaceans such as prawns, and *Octopus* feed on crabs and other slower-moving Crustacea. It would be fascinating to know something of feeding and digestion in deep-water squids, but our knowledge is at present almost wholly of more familiar inshore forms, much of it due to the studies of Dr Anna Bidder on *Loligo* and other genera.[68] There must be many variations in detail, and probably wider differences, too, in other cephalopods. None the less, *Loligo* may be a little more representative of cephalopods than would be any single snail of the wide variety of gastropods.

Most cephalopods are of large size, and the digestive system has achieved speed of action by reducing its reliance on the ciliary and mucous mechanisms of other molluscs. Smooth muscles are now all-important (the gut is for the most part highly muscular), and there is a delicate nervous co-ordination of peristaltic movements, the opening and closing of sphincters and the activity of glands. The splanchnic ganglion, lying alongside the stomach, is a part of the sympathetic nervous system that has not called for special notice in other molluscs. In cephalopods it is responsible for the well-integrated rhythms of the complicated stomach and digestive gland. The action of the buccal mass and the salivary glands is also under nervous control. Slow intracellular digestion is no longer found.

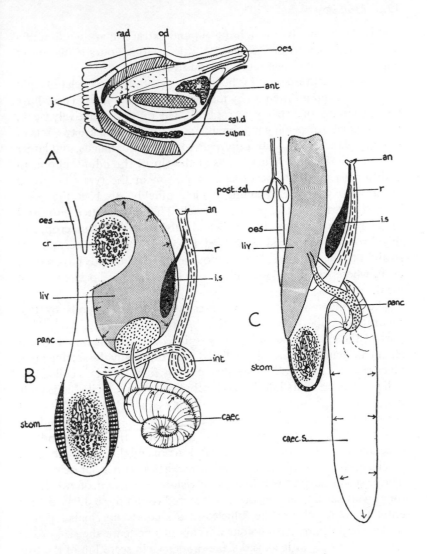

Figure 39 The cephalopod digestive system (*after Bidder*)
(A) Longitudinal section of buccal mass of *Sepia*; digestive tract of
Octopus (B) and *Loligo* (C) (fine arrows indicate absorption sites)

an, anus; ant, anterior salivary gland; caec, caecum; caec. s, caecal sac;
cr, crop; od, odontophoral cartilage; int, intestine; i.s, ink sac; j, jaws;
liv, 'liver'; oes, oesophagus; panc, 'pancreas'; post.d, duct of posterior
salivary gland; post. sal, posterior salivary gland; r, rectum with mucous
strings of faeces; rad, radula; sal.d, posterior salivary duct; stom,
stomach; subm, submaxillary salivary gland

Food travels rapidly through the gut and extracellular digestion of a heavy meal may in *Loligo* take as little as four hours. Cilia are of importance only in one region of the stomach.

In *Loligo*, captured food is held at the mouth by the circlet of eight short arms, and is bitten up by the two overlapping jaws. These have the appearance of a strong parrot beak. The radula is relatively weak, far too small for rasping and breaking up food. There are two sets of salivary glands, a small anterior pair which secrete mucus, and larger posterior ones whose duct opens at the tip of the odontophore. In addition to mucus these produce a poison of the tyramine group (parahydroxyphenylethylamine) which quickly disables the prey. They secrete a powerful protease as well. The food is bitten into small pieces before ingestion. In squids and cuttlefish there is no crop, and this broken-down food is carried through a narrow oesophagus straight to the stomach. The octopods and the nautiloids have a deep crop, where fragments of crab meat or other food are packed after a meal.

The molluscan stomach is represented in cephalopods by two separate chambers communicating by a sphincter (Fig. 39B). That part of the early stomach which was lined with cuticle now forms a strong muscular gizzard. The ciliated part, especially the sorting area, gives rise to a thin-walled caecum. In many families this is coiled in a helical spiral, and its interior is closely lined with tall ciliated leaflets which converge on the intestine. In many squids, such as *Loligo*, it is prolonged into a thin-walled tapered *caecal sac* which extends to the end of the visceral mass. At the junction of the gizzard and the caecum the ducts of the digestive glands open and the intestine begins.

The cephalopod digestive gland also exists as two separate organs. The first and largest is a long yellowish-brown gland of two fused lobes with paired, narrow ducts. Clustered round these ducts as they enter the stomach are the follicles of a second and smaller gland, wedge-shaped, cream in colour and about one-tenth the size of the larger. We know little as yet of the difference in secretion of the two glands or their separate role in digestion. The outflow of the one is brown, of the other colourless. Until we can use more precise names, we may retain the old terms 'liver' and 'pancreas' respectively.

The different functions of the stomach are carefully regulated. In *Loligo* food goes first from the oesophagus to the gizzard, where its mechanical breakdown is completed. It is mixed here with 'pancreatic' fluid which has flowed in by the opening of the sphincter leading to the caecum. Partly digested food is from time to time released into the caecum as the sphincter relaxes, and there it receives the 'liver' secretion.

For a while the two secretions are confined to separate chambers, and a separate stage of digestion may go on in each. As food is gradually yielded from the gizzard to the caecum the ciliated leaflets efficiently remove indigestible solid fragments, which are carried to the intestine. This recalls the sorting activity of other molluscan stomachs, though there is in *Loligo* no protostyle rotating in the intestine, and all other movements of food are performed by muscle. The smooth part of the caecum, including the long caecal sac, is an absorbing area, and absorption later spreads to the intestine. Last of all, the valve between the gizzard and intestine opens and a compact bolus of faeces enters, made up of the last remains left in the gizzard.

Nautilus and octopods differ from decapods in having a very distensible crop attached to the oesophagus, which is packed with bitten-up food after a meal. In contrast with *Loligo*, digestion of a meal in *Octopus vulgaris* may take twelve to fourteen hours. Here the liver is the absorptive organ, and since its phases of secretion alternate with those of absorption, assimilation of the crop contents takes place in several instalments. In *Sepia officinalis*, too, Bidder finds the liver to absorb, with digestion prolonged up to eighteen to twenty-four hours.[68]

In the second volume of the *Physiology of Mollusca*, Dr Bidder has given an excellent account of the digestion of Cephalopoda, incorporating much original work on the structure of the gut and its distribution of functions in *Nautilus*, decapods and octopods.

The cephalopod gut has achieved many of the advances found in the higher chordates. This is especially true of the perfection of nervous control, the speed of digestion attained in *Loligo*, the reduced importance of cilia and the loss of intracellular digestion. The radula is now insignificant compared with the jaws; and the chief ciliated area that remains has a new use – rapid clearance of waste after digestion. The essential pattern of the molluscan gut, however, remains; some of its features are very stable, but – as Bidder has shown – 'the cephalopods have been able to use the molluscan inheritance to form a digestive system of startling efficiency'.

6   Blood, body cavity and excretion

In all molluscs but cephalopods the general body cavity is filled with blood brought back from outlying venous sinuses. The true coelom comprises only the pericardium and the cavity of the gonad. In chitons, gastropods and lamellibranchs the head, foot and viscera are supplied directly with blood by closed arteries from the ventricle. There are in molluscs no true capillaries except in the advanced and well-endowed cephalopods; in other classes blood seeps from the arteries into pseudovascular spaces in the connective tissue. In gastropods, peripheral blood from the head, foot and much of the mantle is returned into a central space known as the *cephalopedal sinus*. From the visceral mass it is brought back to a *visceral sinus*. Both these spaces discharge into a *subrenal sinus* lying near the columellar muscle at the base of the visceral mass. From the subrenal sinus blood is distributed in varying proportions through the respiratory organs and the kidney before returning to the heart. There is an extensive *renal portal system* into which all or part of the venous blood may go, passing thence either to the gill or directly to the auricle. By an alternative route, the *rectal sinus*, blood may be sent along the mantle roof straight to the gill without deploying through the kidney. Re-oxygenated blood always returns by the efferent branchial vein directly to the auricle.

Several memoirs[2, 16, 18] describe in fine detail the arrangement of the molluscan vascular system. In this book we can only – by the diagrams in Fig. 40 – show the general course of the circulation in a diotocardian and a monotocardian prosobranch, examples not too untypical in their broad features of the Gastropoda at large.

Some molluscs lack a heart. In the Scaphopoda there are no well-defined blood vessels of any sort, and the pericardium is also lost, though paired kidneys remain. Blood circulates between the organs largely by contractions of the body wall, and is re-oxygenated in simple transverse folds of the mantle roof. In some Sacoglossa, among gastropods, the blood is pumped not by a heart but by the

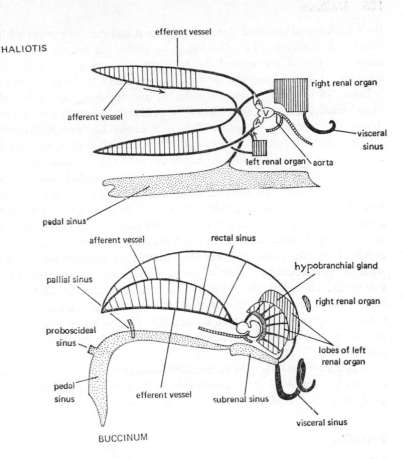

HALIOTIS

efferent vessel

afferent vessel

right renal organ

visceral sinus

left renal organ · aorta

pedal sinus

afferent vessel

rectal sinus

pallial sinus

hypobranchial gland

right renal organ

proboscideal sinus

pedal sinus

efferent vessel

subrenal sinus

lobes of left renal organ

visceral sinus

BUCCINUM

Figure 40 Schematic plan of circulatory system in a diotocardian, *Haliotis* and a monotocardian, *Buccinum*

muscles of the body wall, especially – in *Alderia* and *Stiliger* – by contractions of the club-shaped cerata.[120] Even when present, the heart may play but a minor part in the shifting of blood: changes in the shape of the body redistribute blood on a scale impossible for the ventricle, and in many gastropods and lamellibranchs the aorta as it leaves the heart is valved against forcible backflow through the arteries.

Cephalopods are much larger than other molluscs and have a much faster tempo of life. As well as the arteries of other molluscs their high metabolic rate demands a closed circulatory system with veins and capillaries. Locomotion is by pallial jet propulsion; and arms and funnel are moved by interplay of complicated intrinsic muscles. It is no longer necessary nor possible to lock up large volumes of blood as a haemoskeleton, and the body space is not blood-filled but forms a true coelom, evidently developed independently for cephalopod needs. This new cavity has an endothelial lining and communicates widely with the gonadial coelom. The heart lies within it, and it is in fact an extension of the pericardium around the rest of the viscera. From the coelom renopericardial ducts open to the kidney near the external renal orifices, and in *Nautilus* these have lost their connection with the kidney to open directly to the exterior. In the Octopoda, for reasons still uncertain, the pericardial coelom is again reduced and the heart excluded from it. It forms only a pair of capsules round the branchial hearts, and these communicate with the gonocoele by a pair of long narrow ducts which also open externally. These have been incorrectly called 'aquiferous ducts' and the coelom a 'water vascular system'. It is possible that such external coelomic openings may serve to equalize the coelomic and pallial pressures during the forcible contraction of the mantle cavity in locomotion.

In primitive molluscs the coelom was undoubtedly more extensive, as may be seen with the perivisceral space of chitons, and the relatively large true coelom now known in *Neopilina*.

Excretion

Like many coelomic surfaces, the molluscan pericardium takes on excretory functions. Two types of excretory organ are developed. First, the pericardial epithelium may itself be thickened to form deep brown-coloured pericardial glands. In lower gastropods and in many lamellibranchs these usually lie over the auricle wall. In higher gastropods the glands are on the pericardial side-wall, and in some lamellibranchs (e.g. Unionidae) they may form extensions leading from the pericardium into the tissues of the mantle, known as Keber's organs. In cephalopods they constitute the glandular appendages of the accessory branchial hearts.

The coelomoducts which open by renopericardial apertures from the pericardium are usually glandular, forming a renal organ or kidney, extracting nitrogenous waste from the blood which is supplied by the renal portal system. Molluscan kidneys have various forms. In chitons they are paired symmetrical tubes, each prolonged forward from its pericardial opening near the renal pore, and provided with numerous small dendritic outgrowths ramifying along its whole length. In Gastropoda the kidney is usually a thick-walled sac, much expanded and massively folded within. In higher Gastropoda only the left post-torsional kidney remains. In Heteropoda and some Gymnosomata the kidney is thin-walled and transparent, while in the dorid nudibranchs it is greatly subdivided into slender branches ramifying among the viscera. In the Prosobranchia the kidney opens into the hinder part of the mantle cavity, in higher Opisthobranchia directly to the exterior on the right side; in the Pulmonata its duct is narrowly drawn out alongside the rectum to open with the anus outside the mantle cavity. The kidneys of bivalves are paired tubes, each bent on itself as a ⊃; the lower (or proximal) limb is glandular, and opens from the pericardium, while the upper (or distal) limb is a thin-walled bladder opening into the mantle cavity. In the primitive Protobranchia the whole tube is excretory. In the Cephalopoda the kidneys are inflated sacs with smooth outside walls. Their glandular tissue had shifted to form the spongy mass of renal appendages surrounding the afferent branchial veins which pursue their course through the middle of either kidney.

The blood in the heart has a high hydrostatic pressure. From it, through the walls of the auricles and ventricle, is continuously transfused a clear filtrate into the pericardium. This fluid is isotonic with the blood but with much less non-mineral matter. With the secretions of the pericardial glands, it passes through the renopericardial openings to the kidneys. Here the products of nitrogenous excretion are added to it by the action of the lining cells. Inorganic substances may be taken back into the blood, and the composition of the resulting urine is determined by the extent of renal secretion and of reabsorption of ions.[254] In freshwater lamellibranchs such as *Anodonta* and in gastropods like *Lymnaea* – where retention of inorganic ions is important – most of the salts are resorbed in the kidneys so that the urine becomes hypotonic to the blood and pericardial filtrate. And even in marine molluscs some amount of ionic regulation exists, resulting in the conservation of physiologically valuable substances. Thus, in the blood of the marine gastropods *Buccinum, Neptunea* and *Pleurobranchus* calcium and potassium exceed the equilibrium values with sea water. There is a slight elimination of magnesium, while sodium and chlorine are in almost identical concentration with that of sea water. In the active bivalves *Pecten* and *Ensis* there is a similar picture; but there is less accumulation of potassium in the more sedentary *Mya*. In the very active cephalopods regulation extends to all the ions, potassium and calcium being much in excess of their seawater value, magnesium and chlorine only slightly above, and sodium and sulphate rather below. Molluscs appear to take up inorganic ions from sea water by absorption through the gill and exposed body surface rather than by the gut. At least with potassium, magnesium and chloride, and usually with calcium, this can take place against a concentration gradient by performance of osmotic work.[273]

Though capable of ionic regulation, marine molluscs are in general permeable and in total osmotic equilibrium with the sea water.[2] They have little power of osmotic regulation when placed in lowered salinities, though over long periods they may osmotically adjust their blood concentration to more dilute media. Many marine molluscs such as *Patella, Nucella, Cardium, Ostrea* and *Mytilus* have been reported to acclimatize to very low salinities. Most bivalves such as *Mytilus* and *Ostrea* prevent loss of salts over short periods in freshwater by tightly closing the shell. The marine *Scrobicularia plana* can live in brackish estuaries; its internal osmotic pressure varies with that of the medium over a wide range, though after an equilibration period there is some suggestion of active control of blood concentration.

Freshwater gastropods and bivalves work with a very low ion concentration. *Anodonta* has indeed the lowest blood concentration recorded for any animal (0·08°C, equivalent to 4–5% sea water). The problem of continuous osmotic inflow of water is solved in two ways. They bale themselves out continually, by producing a copious urine, about 25% of shell/body weight daily, or 65% of the extracellular fluid. This urine is strongly hypo-osmotic with the recovery of salts by the renal organ. In addition there is hyper-osmotic regulation, especially with the uptake of ions such as Na^+ and Cl^- at the body surface. Dilute urine can reduce the osmoregulatory requirement by 80–90% but in *Anodonta* this still consumes 1·2% of the total metabolic energy.[254]

In land pulmonates water must be no longer baled out but conserved. Here the renopericardial aperture is very small and little if any fluid is sacrificed from the pericardium to the kidney. A form of nitrogenous excretion is evolved which requires little water. Like birds and reptiles, land gastropods excrete almost insoluble uric acid. They are said to be *uricotelic* and white crystalline deposits accumulate in the gland cells of the kidney, being discharged as spherules at rather long intervals – four to six weeks in the Roman snail – or remaining during hibernation in a 'kidney of accumulation'.

Much recent work has been carried out on prosobranch excretion, reviewed – up to 1967 – by Potts' important survey of molluscan excretion,[258] Subsequent studies by Little on *Strombus*[195] and neritaceans, and by Webber and Dehnel on *Acmaea*[303] shows that these marine prosobranchs do not regulate their haemolymph to any extent, and that the kidney is not involved in output or uptake of ions, though it may both secrete and resorb certain organic substances.

The Architaenioglossa, early prosobranchs to depart from the sea, have been carefully studied by Andrews and Little,[50] with details of renal ultrastructure. The family Viviparidae live in fresh water and differ only from marine prosobranchs in having a pallial ureter that can absorb both salts and water. The Ampullariidae are freshwater snails that can aestivate in dry periods, as can also the Cyclophoridae, that are specialized for fully terrestrial life, though still in damp places. The kidney in these two families is subdivided into two chambers. The posterior one, the 'old' kidney, is concerned both with excretion and with storage and resorption of water during aestivation.

An anterior chamber has developed by the forward growth of the kidney into the mantle roof: its walls are folded and specialized for salt resorption (and also secrete yellowish non-uric particles).

In the terrestrial pulmonates, the role of the pericardium in the filtration of urine is totally lost. In the renal sac there are two processes, the elaboration and extrusion of crystalline concretions of uric acid, and the ultrafiltration of haemolymph, with the production of a liquid urine, isotonic with it. In the primary ureter, there is resorption of water and of salts, with the resultant 'secondary urine' hypotonic to the haemolymph. The secondary ureter modifies the urine by further resorption of water across its wall; this is complete in dry habitats, but decreases in importance in habitats where moisture is more abundant. In Stylommatophora, the ureter thus appears to be an essential organ of water and electrolyte regulation; but in limnic Basommatophora it is found to remain in simple tube.

Most aquatic pulmonates excrete ammonia, either directly or in part converted to amines and urea. Ammonia is a poisonous substance requiring dilution in a large volume of water and the freshwater bivalve *Anodonta* appears partly ammonotelic. Bivalves, in fact, being thoroughly aquatic, seem rarely to form uric acid at all. Needham found the nitrogenous excreta of *Mya arenaria* distributed

E

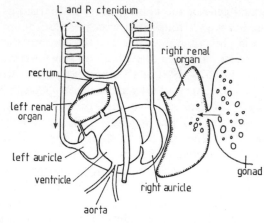

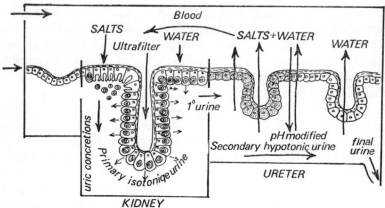

Figure 41
Above Renal-pericardial complex of *Haliotis*, a diotocardian prosobranch
Below Scheme of excretion in the kidney of *Helix pomatia* and *Archachatina ventricosa* (*after Vorwohl*)

as 21·5 parts ammonia, 4·5 parts urea and 18·0 parts amino acids and creatine. The renal sac fluid of cephalopods yields a very high proportion of ammonia, sometimes one-third to two-thirds of the nonprotein nitrogen, with smaller amounts of purines, amines, urea and sometimes uric acid. There is never the high rate of conversion of ammonia to urea seen, for example, in mammals. In the water of a cephalopod aquarium there may be even a higher proportion of ammonia to other nitrogenous matter than in the urine; additional ammonia is thus probably excreted by the gills. As well as the pericardial glands and kidneys, the cephalopod digestive gland is an important excretory organ. In 100 g of fresh 'liver' of *Octopus* were detected 30–50 mg of ammonia, 6–25 mg of urea as well as 3–17 g of uric acid. As in the vertebrate liver, excretory products may be primarily formed in the digestive gland. They are extracted from the blood by the pericardial glands, renal organs and gills.[273]

Blood

The respiratory pigment of Gastropoda and Cephalopoda is haemocyanin, a copper-containing compound carried in the plasma, which it colours faintly blue. This has similar oxygen-carrying properties to haemoglobin, though different molluscs vary widely in the oxygen pressures at which their haemocyanin becomes saturated. Thus the bloods of most gastropods are saturated at low oxygen pressures, a factor fitting many Gastropoda to occupy habitats of poor aeration. On the other hand, cephalopod haemocyanin becomes saturated only at relatively high oxygen pressures, and these active molluscs are extremely sensitive to oxygen lack. Dissolved oxygen never forms more than about 3% of the total blood volume in molluscs, except in cephalopods, where the percentage is 8 to 11, as compared with 20% in mammals.

A few gastropods that live in especially poor oxygen conditions have developed haemoglobin, which is here capable of working at very low oxygen pressures. In *Planorbis corneus* living in foul muds, for example, the haemoglobin is saturated in all parts of the body and thus useless, in a medium containing more than 7% dissolved oxygen. With pressures from 7% down to 1% (where the animal dies), haemoglobin is useful; it is never so in molluscs for aerial respiration. The fast-working muscles of the gastropod buccal mass frequently contain muscle haemoglobin, which may possibly work as an oxygen store during muscle contraction.

It is not certain that any lamellibranch has haemocyanin, and in most species studied there is no oxygen carrier in the internal medium, the blood oxygen concentration being that of the outside water. Such a mechanism is able to supply sufficient oxygen to a sedentary animal by reason of the ample surfaces for oxygen uptake in the enlarged ctenidia and mantle. Some bivalves, particularly those living in poorly aerated substrata, possess haemoglobin, contained in non-amoeboid corpuscles. Examples include many of the Arcidae, and *Solen legumen* living in muddy sand in the Mediterranean.

7　Sex and reproduction

At the beginning the molluscs had no separate genital ducts. Paired or single gonads opened into the pericardium and the coelomoducts (renal organs) carried eggs and sperm directly to the sea, where fertilization normally took place. Such an archaic state persists today in some Aplacophora and – as we now know – in *Neopilina*; and the gametes still pass through both kidneys in early bivalves, and through one in Archaeogastropoda and Scaphopoda. With few exceptions fertilization is external.

The chitons – which are otherwise primitive – have acquired separate ducts leading from the gonocoele and opening by distinct genital pores. The gonad is single and median, and the sexes are separate. Most species produce a large number of eggs which they may shed freely or – sometimes in the same species – enclosed in short mucous strings secreted by the gonoduct. In *Lepidochitona cinereus* the eggs have a protein membrane, like a ruff or pie-frill, produced by the follicle cells of the ovary. The larva is a modified trochophore spending only about six hours in the plankton and having sufficient yolk for its needs without planktonic food.

The Caudofoveata have separate sexes and a single gonad. In the Ventroplicida, which are hermaphrodite, the gonad is paired, and some early reproductive specializations appear: the renal organs, through which the gametes pass, are equipped with simple egg-shell glands and sperm receptacles and there are protrusible cloacal spicules which seem to be copulatory organs.

In primitive gastropods – as in the earliest molluscs – the eggs have very little yolk, and – being fertilized externally – can have no thick reproductive capsule. The first larva is a freely swimming trochophore, followed after a few hours by a veliger. Such molluscs are mostly restricted to shallow inshore seas. The life history is a critical barrier to new habitats: external fertilization in fresh water is impracticable, for free sperms are usually unable to resist lowered salinity. Thus the Archaeogastropoda, and the chitons too, are for the most part marine and coastal. A few groups only, like the surviving Pleurotomariidae, and some lepidopleurids, have gone into deeper waters.

Gastropods of whatever kind have but a single gonad, and all the Archaeogastropoda, except the specialized Neritacea, use one of the renal organs, i.e. the post-torsional right kidney, for the passage of gametes. The gonad may open directly into the kidney (*Patella* and *Haliotis*) (Fig. 41), into its duct, or as in trochids into the reno-pericardial duct. In the higher prosobranchs the renal function of this kidney is quite lost and it then survives incorporated as a short section of the genital duct; it may still retain a narrow pericardial connection, the *genitopericardial duct*.

The genital products in most Archaeogastropoda are thus discharged far back in the mantle cavity. Except in Neritacea there is no penis, and hardly ever a secretory oviduct. The sexes can be distinguished only by the colour of the gonad and the sort of gametes it produces.

Limpets like *Acmea* and *Patella* shed their eggs singly into the plankton. They have no protective covering – only a thin membrane and an albumen layer which is soon lost. In *Haliotis*, as in most trochids, and in *Tricolia pullus*, the eggs are also shed singly but are further surrounded by a thin gelatinous coating. In the Fissurellidae – such as *Diodora* – these gelatinous sheaths are joined to each other so that the spawn is deposited in a coherent layer. A few higher trochids, such as *Cantharidus* and *Calliostoma*, have progressed further. They form a gelatinous egg ribbon attached to the substrate and secreted by the lips of the renal oviduct or – in *Calliostoma zizyphinum* – by part of the female's mantle wall.

Even in archaeogastropods a free-swimming larva is not always found. Only nine out of the seventeen British species have trochophores; in the rest – including *Diodora* and *Calliostoma* – the veliger hatches from the egg membrane at the creeping stage. Larval life is frequently short. White *Patella* spends ten days in the plankton, *Haliotis* is limited to about forty hours, and torsion is not finished until after settlement.[95]

In those molluscs that shed their eggs and sperm into the plankton – this includes some archaeogastropods and nearly all marine lamellibranchs – great wastage might be expected from failure of fertilization. Many species, however, have adjusted their spawning behaviour to permit economy of gametes. Thus some species of *Patella*, as well as at least one *Helcion* and *Gibbula*, will not shed their sex products unless close to one of the other sex. In many molluscs, too, the males spawn first and shedding of eggs is only induced in the presence of sperms (some chitons, *Haliotis* and *Ostrea*). In still other forms, particularly lamellibranchs, there is 'epidemic spawning', perhaps controlled by temperature and food abundance, or showing lunar periodicity as in *Pecten opercularis*.

Further evolution in gastropods demanded special genital ducts (Fig. 42), first to convey and to receive sperm at internal fertilization, and then to provide better nutritive and protective layers for the eggs. These were prerequisites for shortening the larval life, or for retaining and brooding the embryos. With this achieved many new habitats were opened up, especially on land and in fresh water. Both these habitats were occupied by the first mesogastropods: it is probably not a coincidence that the Neritacea, the only archaeogastropods to develop an oviduct and a penis, have at once run through the whole programme of later pulmonate evolution. They have spread to rivers and freshwater lakes, and to tropical rain forests.

Both sexes obtained a glandular genital duct by pressing into service part of the right side of the mantle. The genital opening was first brought forward by an open furrow as far as the front of the mantle cavity. In the female this soon became closed to give a glandular tube by which the eggs were carried from the original genital opening to the mantle edge. Glands were differentiated in the order in which secretions were placed round the egg – first appears a coat of albumen, then a tough capsule, and finally – as in some littorinids – a mass of jelly. Thus appeared in sequence an albumen gland, capsule gland and sometimes a jelly gland. In addition, now that the new opening could be reached by the male's penis, a ciliated channel ran up the female duct to a receptaculum seminis, a pouch inserted between the albumen and capsule glands (Fig. 42B). From the new female aperture, near where the mantle joined the body wall, a ciliated groove ran forward along the head and foot to the ground level. By this means the fertilized eggs could travel forward and be fixed against the substrate.[129]

In most mesogastropods an elaborate spawn is formed, the young hatching either as veligers or at the crawling stage. There is a wealth of information on egg masses and larvae to be found in the papers of Lebour and Thorson and Fretter. In some of the littorinids and other early families the separate capsules are laid in an egg mass of jelly (*Littorina obtusata* and *Lacuna*). *Littorina littorea* and *L. neritoides* on the other hand have planktonic egg capsules and veligers, while *L rudis* is viviparous. In other mesogastropods groups of several eggs are usually enclosed in thick capsules. In the Rissoidae these are lens-shaped or spherical and attached to the substrate. In the Turritellidae they are fastened in grape-like clusters. *Cerithiopsis* places them in nests in sponges. The Lamellariidae and Triviidae plant vase-shaped capsules in the tests of the compound ascidians on which they feed. The Calyptraeidae protect the thin capsules under the parent shell. In the Naticidae the egg capsules are glued together with sand, to form the characteristic smooth ropes or egg collars that encircle the animal as they are secreted. Some species of pelagic *Ianthina* carry the egg capsules attached to their raft.

Land operculates such as the Cyclophoridae have perforce lost the larvae altogether and deposit yolky eggs in tough envelopes. Most freshwater families such as the Hydrobiidae, Melaniidae and Valvatidae prefer to brood the young, rather than entrust veligers to the hazards of running streams. Incubation also takes place in many marine families and the types of brood pouch are very diverse. Thus the Siliquariidae, like some Ianthinidae, and the freshwater *Tanganyicia*, have a spacious pouch in the head, opening beneath the right tentacle. The Struthiolariidae have a brood pouch in the mantle. *Viviparus* uses the oviduct as a uterus. The siliquariid *Stephopoma* carries the young freely in the mantle cavity, while the Vermetidae attach a row of capsules to the inside of the mother's shell.

The largest and finest veligers are those of Mesogastropoda. In three families, the Lamellariidae, Cypraeidae (Eratoinae) and Capulidae, the veliger is an *echinospira* with the young definitive shell covered by a much larger secondary shell, the *scaphoconcha*, which has practically no weight and assists in flotation during the long swimming life.

A few Neogastropoda, as the Nassariidae and Turridae, have also free larvae, but for the most part their eggs are placed in horny capsules, the majority serving as nurse eggs for the few cannibalistic survivors that hatch at the crawling stage. The Buccinidae and their relatives produce an elaborately compacted spawn mass, but in the Muricacea the egg capsules are separate, and may be vase-shaped (*Nucella*), lens-shaped (*Ocenebra*), or cylindrical as in many species of *Thais*. Many of the Volutidae form spherical calcareous capsules attached to other shells. A great difference between mesogastropods and neogastropods is that in the latter the female employs her foot, which develops a ventral pedal gland, to mould the unfinished horny capsule that emerges from the capsule gland. Fretter has given a detailed account of the way the British *Buccinum, Ocenebra, Nassarius* and *Nucella* manipulate these capsules with the foot and attach them to the substrate.

In male mesogastropods and neogastropods the glandular duct forms a so-called 'prostate'. Mixed with its secretion the sperms are carried from the mantle cavity to the head by a ciliated groove, which may sink in to form a shallow vas deferens. The foot develops a muscular penis attached below the right tentacle. This is grooved by the male genital furrow and kept reflected into the mantle cavity when out of use. The tropical freshwater snails of the Pilidae uniquely develop a penis on the mantle edge.

In higher prosobranchs fertilization is always internal, but there are some species where this cannot be effected by copulation. First there are sessile forms like Vermetidae and Hipponicidae which may be out of reach of other individuals. In others, such as the Turritellidae, the mantle cavity is kept closed by a portcullis of pinnate tentacles against the entry of sediment – and thus also of the penis. Thirdly, in many tightly coiled or long-spired gastropods, such as *Bittium, Cerithiopsis, Clathrus* and *Cerithium*, the mantle cavity is very narrow and the female opening is too far up the spire for a penis to reach it. In all of these forms the male is aphallic and sperms are shed freely into the water and enter the mantle cavity of the female with the inhalant current.[129] The genital duct is split like an open sleeve for entry of sperm, but how they finally reach it is a mystery. There is no authenticated case in animals of sperms being guided by chemotaxis.

The strangest sexual phenomenon is the presence of two types of sperms in many families. The most numerous are small and normal *eupyrenic* sperms which fertilize the eggs. The others are much larger, reaching sometimes 100μm in length. They sometimes lack flagella and swim with an undulating membrane like sphirochaetes. In *Viviparus* they bear a tuft of flagella at one end. The nuclei are degenerate and these *oligopyrenic* sperms evidently play no part in fertilization. The usual view is that they are nurse cells which disintegrate in the receptaculum to nourish the eupyrenes. But, as Ankel has shown, they can also have a transporting function. Huge numbers of normal sperms may attach by their heads to a vermiform sperm. The whole structure – known as a *spermatozeugma* – can swim strongly in sea water, and in aphallate genera – as *Clathrus* and *Ianthina* and probably *Cerithiopsis* – fertilization is achieved by the active entry of the carrier-sperm into the oviduct.[131]

Sex and hermaphroditism

It has been widely believed that primitive gastropods are bisexual, and that hermaphroditism is the hallmark of opisthobranchs and pulmonates. In 1909 Orton made his famous discovery that the slipper limpet, *Crepidula fornicata*, begins life as a male, passes through an hermaphrodite stage and finishes as a pure female. It was gradually realized that in more and more groups scattered among the prosobranchs the sexes may – for a part of the life – be united in one individual. Sometimes, perhaps always, this hermaphroditism is of a protandric kind. Coe has called this *protandrous consecutive sexuality*, and it is perhaps best studied in the chains of sexual individuals formed by *Crepidula fornicata*. The first and youngest individual is a tiny male at the summit of the chain. It lies in the mating position on the right side of the shell of the young female lying next beneath it. The male has a large testis and a muscular penis which conveys the sperm. In older slipper limpets, lower in the chain, a hermaphrodite stage follows. Oocytes develop in the gonad, sperms cease to form, and the penis is partly absorbed. The oldest specimens, at the bottom of the chain, are mature females. The ovary and oviduct are functional, the penis is almost lost, and a brood of spat lies under the shell. Shorter chains – or associated pairs, small male and large female – are found in other Calyptraeidae, such as *Calyptraea*, *Janacus* and *Crucibulum*. With the change to female the animal becomes more sedentary, and merely broods over the eggs. Sex change can be modified by temperature, by food and by the influence of association. By removal of the female a young male can be induced to pass sooner into the neutral or female phase.

Consecutive sexuality – with protandry – has been found by Ankel in *Ianthina* and *Clathrus*. In the freshwater *Valvata tricarinata* there is a more complicated *rhythmical sexuality*, with a regular return to a male phase after the eggs are laid. The list of hermaphrodite forms – whether or not they are known to be protandric – includes some Hydrobiidae, the Omalogyridae, *Velutina* and *Puncturella*. Other hermaphrodites are the parasitic forms included in the Eulimidae and Entoconchidae, though some of the latter have minute dwarf males (see p. 196). A moderate male dwarfism occurs in *Lacuna pallidula*, where the male is one-tenth the weight of the female and lives attached to her shell.

Orton and his colleagues have also investigated the sexuality of the limpets.[244] In *Patella vulgata* 70·8% of the smallest individuals measured were males and only 4·4% females. Of the oldest age group 34·7% were males and 64·2% females. In Italy, Bacci has had similar results with *Patella caerulea* and *Fissurella nubecula*. This might suggest that the females grow larger than the males, but it also points to a change of sex within the lifetime of the limpet, and, as a further hypothesis, Orton considered that there might be two types of male limpets, true males and temporary (protandric) ones. Most recently Dodd has further examined Orton's material and has in *Patella vulgata* found only a minute fraction of hermaphrodites (30 individuals in 60000).[244] Among other limpets, however, Thorson has shown that *Acmea rubella* is hermaphrodite and *ambisexual* – that is, with both ova and sperms produced side by side.

In Polyplacophora and Caudofoveata the sexes are said to be always separate, though a detailed study of sex in a chiton, along the lines of the limpet work, would be of the greatest interest. The Ventroplicida are protandric, with a very short male phase.[289]

All opisthobranchs and pulmonates are hermaphroditic. In higher forms of both groups – such as nudibranchs and helicid snails – eggs and sperm seem to mature simultaneously. In early members, consecutive sexuality with protandry is prevalent, a brief male phase preceding a mixed or pure female stage, as in – for example – the pteropods (Limacinidae, Clionidae) in the opisthobranchs and the Ellobiidae and *Otina* among pulmonates, which show a protandrous succession in a single season.[221] In the terrestrial *Carychium* there is some evidence that a second male phase follows egg-laying. Other established protandric genera are the pulmonates *Physa* and *Limax*, and the opisthobranchs *Aplysia* and *Aeolidia*.

Evidence from many groups of gastropods indicates an early, deep-seated tendency to protandry, with the sperms – metabolically less expensive – produced earlier in the season or by a younger animal. Later prosobranchs have evolved towards separate sexes; the higher opisthobranchs and pulmonates on the other hand have developed simultaneous hermaphroditism.

True parthenogenesis is very rare in molluscs. In two gastropods, however, *Campeloma rufum* and *Hydrobia jenkinsi*, males are unknown.

Higher gastropods

Beginning from the simple pallial genital duct of prosobranchs the Opisthobranchia have developed very complex genitalia. The male and female portions separately represented in prosobranchs are here united in one individual. The capsule gland has been replaced by a spacious mucous gland secreting a jelly-like egg-mass or ribbon. This gland, together with the albumen gland, sperm receptacles and prostate, has sunk deeply into the general body cavity; and each gland may be elaborately subdivided, or cut off from the main path as one or more diverticula. The higher opisthobranchs – such as the Aeolidiacea and Doridacea – reserve a central part of the gonad for the formation of sperms, with a series of outlying follicles producing the eggs.

Early opisthobranchs have a ciliated seminal groove as in proso-branchs running forward from the common genital aperture to the penis (Fig. 42D). With the loss of the mantle cavity to contain it, the penis has become invaginated within the head, and in later forms the common aperture moves forward to lie near it on the right side of the head, so that both male and female ducts open together.

In most nudibranchs each partner fertilizes the other. Aplysioids, on the other hand, such as *Aplysia* and *Akera*, form a copulatory chain in which each animal acts as a male to the one in front and a female to the one behind. The arrangement of the female ducts becomes very complicated indeed in the Doridacea (Fig. 42F). Here the vaginal section of the female duct separates from the oviduct to give a third genital aperture, the old female duct being reserved for secreting the egg-mass and discharging the eggs and the new duct forming a copulatory passage. This connects with two sacs for storing sperm, a bursa copulatrix which forms a temporary store after copulation, and – higher up – a receptaculum seminis which is the homologue of the same sac in prosobranchs (Fig. 42B).

The aeoliids and tritoniids retain an undivided vagina, but in some Sacoglossa the copulatory arrangements may be very novel. With *Limapontia*, *Actaeonia* and *Alderia* inward sperm travels by a more direct third passage, formed by the bursa copulatrix whose swollen tip rests close to the body wall (Fig. 42G). This bursa may acquire its own external opening, or – as in *Limapontia capitata* and *Actaeonia cocksi* – the body wall may be ruptured at copulation by a hollow spine, or *style*, on the penis of the partner, and sperm transferred by hypodermic injection.[143] In *Alderia modesta* impregnation is reported to take place through almost any part of the body wall. The whole haemocoele becomes charged with sperm, some of which finds its way to the internally open bursa.[271]

Opisthobranch egg-masses are easily recognized (Fig. 42F, G). The

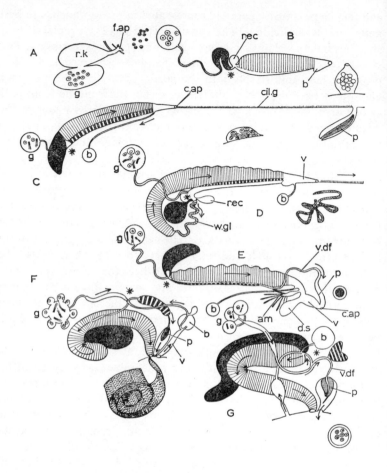

Figure 42 Evolution of the genital duct in Gastropoda

(A) Female *Haliotis* (Archaeogastropoda), with planktonic eggs
(B) Female *Nucella* (Neogastropoda), with egg capsule
(C) *Ovatella* (a primitive pulmonate) with mass of spawn jelly
(D) *Aplysia* (an early opisthobranch), with spawn string
(E) *Helix* (an advanced pulmonate) with shelled egg
(F) *Archidoris* (a nudibranch) with spawn ribbon
(G) *Actaeonia* (a sacoglossan) with egg capsule

In each case, the albumen gland is black, the capsule or mucous gland lightly cross-hatched, the prostate or equivalent male tract heavily cross-hatched

a.m, accessory mucous gland; b, bursa copulatrix; c.ap, common genital eggs are usually small, each with its own albumen coat and egg membrane, and the whole mass is enclosed in a thick matrix of clear mucus. The bulloids such as *Philine* deposit a jelly sphere through which runs a tangled string of eggs. Similar egg-masses of *Actaeon* are club-shaped. The aplysioids – including *Akera* – put out a long mucous string like yarn, which is deposited in a loose tangle. The pleurobranchoids, the dorids and the aeoliids produce a broad flat ribbon, studded with tiny eggs and attached in a coil or in folds to the substrate.

The Opisthobranchia are all marine, and nearly every species has small free-swimming veligers which hatch early and spend only a short time in the plankton. The best-known exception is the high tide pool sacoglossan *Actaeonia cocksi*. Here the eggs are large and yolky, enclosed in a spherical capsule (Fig. 42G), and the larva hatch at the crawling stage. *Alderia modesta* – a sacoglossan of even higher salt-marshes – might be expected to breed in the same way. Paradoxically it lays small eggs and has normal veligers. Like the supratidal *Littorina* (*Melarapha*) *neritoides*, it evidently relies on the opportunities of distribution afforded by larvae. Further, yolky eggs are probably over-expensive to produce in a semi-terrestrial species living near the margin of its range.

The aquatic pulmonates have a hermaphrodite genital tract essentially comparable (Fig. 42C) with that of early opisthobranchs. The pallial genital glands have moved to the haemocoele, and the ciliated seminal groove has sunk to form a vas deferens, running to the penis within its sac in the head. Pulmonate eggs are much more yolky than in most opisthobranchs and free-swimming larvae are released only in such primitive marine forms as *Melampus*. *Amphibola* and *Siphonaria* have capsulate veligers in a jelly-like spawn coil as in opisthobranchs; for the most part – in Basommatophora – the eggs are clustered in a jelly resembling frog spawn, each having its own albumen layer and tough capsule.[111]

In the land pulmonates the eggs and genital system are adapted for new conditions. By comparison with other gastropods the eggs are

---

aperture; cil.g, ciliated seminal groove; d.s, dart-sac; f.ap, female aperture; g, gonad; p, penis; rec, receptaculum seminis; r.k, right kidney; v, vagina; v.df, vas deferens; w.gl, 'winding gland' portion of mucous tract; * fertilization site few in number, extremely yolky and sometimes very large indeed. The largest species of the tropical *Achatina* lay eggs with limy shells, the size of thrushes' eggs. The typical egg shell is the tough elastic one – partly calcified – of garden snails and slugs, and this is laid down as the secretion of the mucous gland. Such shells protect the eggs from small predators but are not impermeable to water, that is to say, the eggs are not, as in land vertebrates, cleidoic. Their cytoplasm is adapted to tolerate considerable water loss rather than their shells to prevent it. Desiccation is also avoided by behavioural adaptations ensuring oviposition in damp and shaded places or beneath the soil.

The genital system of *Helix* (Fig. 42E) is an example often dissected. Most of its new features are adaptations to assist courtship and the successful transfer of sperm, a hazardous operation for a land mollusc. The sperms are enclosed by the male duct in a chitinous envelope, the *spermatophore*, secreted in the *flagellum*, a tubular outgrowth of the penis. Its transfer at copulation is assisted by the lubricating secretion of two clusters of branched mucous glands which open near the mouth of the vagina. The vagina develops also a muscular caecum, the *dart sac*, in which is produced a fine-pointed calcareous shaft, about 10 mm long, and delicately ridged. This is the *telum amoris*, or 'love dart', which is exchanged by the partners with some velocity before courtship, lodging in the integument and serving as a releaser stimulus for courtship behaviour. In some slugs the dart sac is lost as such, and is converted into an exsertile stimulating organ, the sarcobelum. Snails may remain several hours in coitus, and some species of *Limax* entwine themselves together in a sheet of mucus, performing a complicated 'Liebespiel' before pairing.

Passage of sex cells

In various gastropods, but particularly Stylommatophora, sperms are transferred in a long-tailed chitinous envelope, the spermatophore, manufactured in pulmonates in the flagellum. This appears not to protect the enclosed sperm during transfer, so much as to facilitate the passage of the enclosed spermatozoa to the correct part of the oviduct and fertilization site. In *Helix*, as illustrated from the detailed account by Hans Lind (Fig. 43) its structure allows some of the spermatozoa to escape through its tail canal, pass from the stalk of the bursa, and reach the spermatheca by way of the oviduct. Only foreign sperm are stored in the spermatheca; but most of the

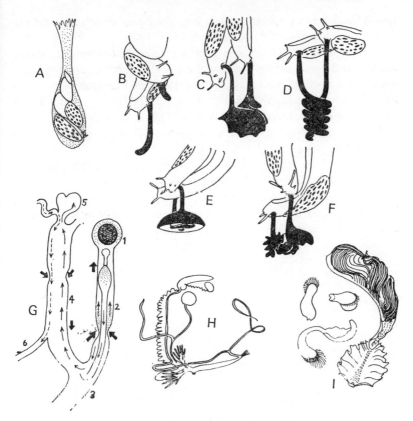

Figure 43 Sperm exchange in Gastropoda

(A–F) The complex aerial courtship of *Limax maximus*
(A) hanging pair entwined in mucus; (B) penial sacs everted and (C), widening to a terminal fan; (D) penial sacs tightly spiralled together; (E) upper coils expanding to form an umbrella shape; (F) penial masses becoming lobed and sperm exchanged while apertures approximated (*based on Chace, 1953*)

(G) *Helix*: sperm migration and transport shortly after copulation, with the spermatophore situated in the bursal stalk (see relations of the parts in (H)). Black arrows at sites of peristalsis; entire arrows show movements of foreign sperm, broken arrows show home sperm. 1 digested excess sperm in bursa; 2 foreign sperm leaving spermatophore body, and 3 migrating down and leaving the tail, and 4 entering the oviduct to reach the spermatheca – 5; 6 home sperm passing down hermaphrodite duct into vas deferens

(I) Spermatozeugma of *Cerithiopsis*, with three stages in development (*after Fretter and Graham*)

sperm received in copulation pass into the bursa copulatrix and are destroyed, as is also the spermatophore.

The diagram shows foreign sperm migrating downwards inside the tail and leaving it to pass up the oviduct. Excess foreign sperm are carried up the bursa stalk by peristalsis and ultimately come to lie in a digested mass within its terminal bulb. The individual's own sperm is however stored in the large coils of the hermaphrodite duct, and is conveyed down the sperm groove of the common sperm-oviduct by peristalsis.

A comparable spermatophore is developed, near the outset of prosobranch evolution, in the Neritacea, where the sexes are separate. The specialization of both male and female tracts, so early acquired, has equipped the Neritacea for their successful essay in both terrestrial and freshwater evolution. In the male, even in marine forms,

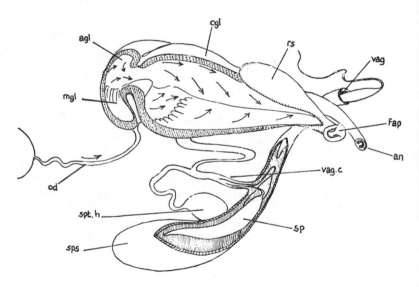

Figure 44 Female genital system of *Nerita melanotragus*
agl, albumen gland; an, anus; cgl, capsule gland; fap, female aperture; mgl, mucous gland; od, ovarian duct; rs, reinforcement sac; sp, spermatophore; vag, vaginal aperture; vag. c, vaginal canal; sps, spermatophore sac; spth, spermatheca

sperms are not shed freely but enclosed in a chitinous spermatophore, transferred by a muscular cephalic penis into a vaginal opening, one of two entirely separate female passages. Sperms escape through the thread-like spermatophore tail, inserted from the storage sac into a 'vaginal canal', that leads into the glandular oviduct and fertilization site (Fig. 44). The eggs when fertilized are enclosed in lidded capsules, secreted and moulded in a capsule gland. In marine species this is studded with calcareous spherules provided from a 're-inforcement sac', and in freshwater *Neritina* by waste particles abstracted from the faeces.

In the hermaphroditic Opisthobranchia a fine account has been given by Beeman of the movements of home and foreign sperm, and ova within the complex passages of the genital tract for

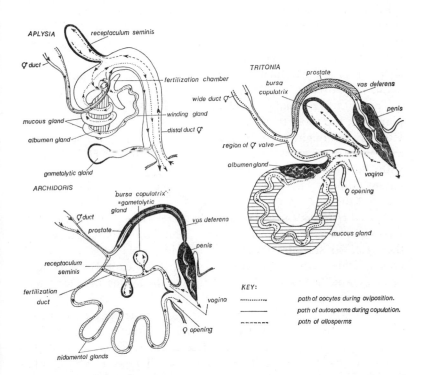

Figure 45 Genital ducts of *Aplysia, Tritonia* and *Archidoris*, showing the pathways followed by oocytes, autosperms and allosperms (*after T. E. Thompson*)

the sea-hare *Phyllaplysia taylori*. Autoradiographic studies of sperm exchange and storage have recently been made.

For the reproductive system, with its varying subdivisions of the female tract, throughout the Opisthobranchia, Ghiselin has published a fine review.[145] Thompson, in his recent monograph, gives much functional detail (see Fig. 45). A similar account for Pulmonata is given by Duncan.[111]

Bivalvia

The reproductive system of bivalves – in contrast with the gastropods – is exceedingly simple (Fig. 46A). The gonads are paired or fused in the middle line. Their ducts are short and have no glands. In early bivalves such as *Nucula* they open into the kidneys and through them to the mantle cavity. In the higher forms they become separate, opening merely on a common papilla with the kidney, or finally through independent pores. Internal fertilization in the strict sense never occurs. In many bivalves, however, the sperms meet the eggs within the mantle cavity, and in two genera, *Xylophaga* and *Cuspidaria*, there are glandular patches near the genital opening that can act as sperm receptacles.[260]

The eggs are usually small and the larvae spend long in the plankton; but those that are incubated in the mantle cavity are provided with enough yolk to nourish the embryo till it is able to settle as a small replica of the adult. There is no 'placental' connection with the mother and the eggs develop morphologically outside the parent's body. The brood chamber is generally the interlamellar space of the inner or outer gill, as in the marine Erycinacea, *Teredo* and *Arca vivipara*. In the larviparous oysters the mantle cavity itself serves as a temporary brood chamber, and free-swimming larvae are then released, spending only a few hours in the plankton. In *Turtonia minuta* Oldfield has described gelatinous egg capsules, unique in bivalves, attached two or three at a time to the byssus and secreted by the glandular edge of the mantle.[241]

All freshwater bivalves appear to incubate the eggs, with the exception of the mussel *Dreissensia*, which is rather recently acclimatized in rivers and canals, and has free veligers.

The freshwater Sphaeriidae liberate small replicas of the adult from between the gills; but in freshwater mussels of the Unionacea the young are released at a much earlier stage, and are incredibly numerous, from several hundred thousand to a million at one time!

They begin life with a phase of parasitism on a freshwater fish, or sometimes a urodele. The larvae are known as *glochidia*, and are equipped with a small triangular-valved shell, and with attachment organs consisting of one or more hinged spines on the shell, or a long byssus thread. In some genera, such as *Unio* and *Anodonta*, the young attach themselves by the shell spines to the fins of the host. Species where the shell is unarmed are attached by the byssus thread to the gills. The glochidia soon become encysted by host tissues, which they proceed to liquefy and assimilate by the margin of their mantle. When the surrounding host tissues are exhausted the larvae undergo histolysis; the adult tissues and organs are then reconstituted, and the metamorphosed young mussel drops off to resume a free existence. European species of *Unio* and *Anodonta* all favour cyprinoid fishes; the American *Lampsilis* may live on the garpike *Lepisosteus* or the bass *Huro*, while *Hemistema* attaches to the gills of the mud-puppy, *Necturus*.

Sex change in lamellibranchs has long attracted interest. While hermaphroditism has been found in only 4% of the species studied, these have been well investigated, and have much theoretic interest. First, there is a group of *functionally ambisexual* bivalves with sperms and ova formed in different regions of the same gonad, as for example in the majority of Pectinidae and in the Tridacnidae, *Sphaerium, Pisidium* and many *Anodonta* species. Or there may be a distinct ovary and testis opening separately on either side, as in members of the order Anomalodesmata, such as *Thracia* and *Pandora*. A second group consists of bivalves having one sex change in the life history, and these are generally protandric, as *Venus mercenaria* (with 98% of individuals being first males) and the wood-boring *Xylophaga dorsalis* and *Bankia setacea*.

The Ostreidae, comprehensively reviewed by Galtsoff[142] and Yonge,[24] show two further sorts of sexuality. The larviparous oysters (e.g. *Ostrea edulis* and *O. lurida*) have *rhythmically consecutive sexuality*. They produce late, short-swimming larvae from the mantle cavity. The youngest gonads are predominantly male, and all individuals begin life as functional males. There follows a series of alternating sex phases, with never a permanent change to female, but always a reversion to male after shedding the eggs. The *Crassostrea* species are oviparous, shedding eggs externally to produce long-swimming veligers. Once thought hermaphrodite, they in fact have a labile balance of alternative sexuality, according to nutritive conditions.

Cephalopoda

In the cephalopods the eggs are comparatively large and yolky, and do not completely cleave, that is to say the embryo is built up from a smaller disc of cells on the upper pole of the egg, and the larger part of the egg goes to form a yolk sac from which the young animal is nourished. In many species the eggs are relatively few, for example sixty in *Eledone*, a hundred or so in *Octopus maorum*. This is not always so: *Argonauta* lays many thousands of eggs at one time, and in the egg-piles of *Loligo*, which are communal, there may be 50000 eggs. The behaviour of cephalopods is much more highly organized than in lower molluscs, and this is particularly so in reproduction. There may be complicated courtship and mating rituals, and very elaborate systems of parental care.

The sexes are always separate. A median ovary or testis lies at the apex of the body and opens straight into the coelom. The oviduct or male duct opens into the mantle cavity at the side of the anus. *Nautilus* has in both sexes a single functional duct (right) and a vestige on the other side. All other male cephalopods, and female Sepioidea, Loliginidae and Cirroteuthidae, have a single (left) duct, while most female Teuthoidea and Octopoda retain the primitive pair. The female duct is very simple; its only appendage is the so-called *oviducal gland* which secretes the coat of albumen round each egg. The outside egg membrane forms an elastic protein tunic, toughening in contact with sea water. In squids and cuttlefish this is secreted by the *nidamentary glands*, of which there are one or two pairs, opening from the ventral body wall near the genital apertures. These pour their secretion over the eggs as they pass from the

---

Figure 46 Bivalvia and Cephalopoda: pericardial and genital organs
(A) Pericardial complex of a typical eulamellibranch
a.a, anterior adductor; au, auricle; g, gonad; g.ap, genital opening; p.a, posterior adductor; per, pericardium; per.g, pericardial gland; r.ap, renal opening; ren, renal organ; r.per, renopericardial opening; rm, rectum; v, ventricle
(B) Male genital ducts of *Loligo* (Needham's sac removed) with (*right*) spermatophore before and after discharge has begun
c, cement body; cm, caecum; cp, cap; ej.d, ejaculatory duct leading into Needham's sac; f.gl, finishing gland forming cap and thread; fil, spiral filament; h.g, hardening gland; i.t, inner tunic; i.t.g, inner tunic gland;

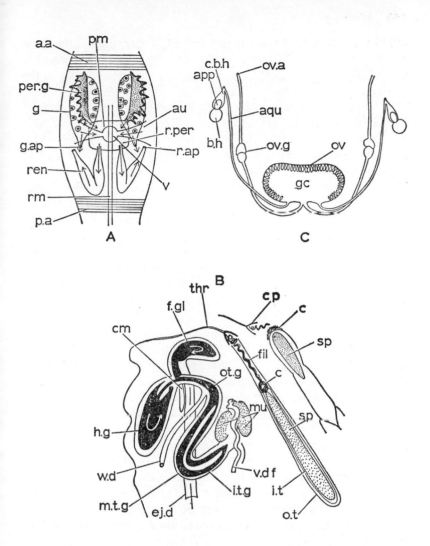

m.t.g, middle tunic gland; mu, mucilaginous glands; o.t, outer tunic;
o.t.g, outer tunic gland; sp, sperm mass; thr, thread; v.df, vas deferens;
w.d, waste duct
(C) Female genital system of *Octopus* (*after Pelseneer*)
app, appendage of branchial heart; aqu, gonopericardial (so-called
'aquiferous') duct; bh, branchial heart; c.b.h, capsule (pericardial) of
branchial heart; gc, gonocoele; ov, ovary; ov.a, oviducal aperture; ov.g,
oviducal gland

oviduct. In *Nautilus* it is the pallial wall itself that supplies the nidamentary secretion. Only in the octopods, where the shell gland is very highly developed, are the egg membranes secreted wholly in the oviduct.

In *Sepia* the eggs are very large (20 mm) and are given a coating of ink as they are shed. They are attached like black grapes to suitable objects on the substrate.[21] *Loligo* encloses the eggs in two or three rows in large sausage-shaped egg-masses, attached at one end in clusters. *Octopus* and *Eledone* fix the single eggs together in clusters. The female *Octopus* expends a great deal of parental care on the eggs, brooding over them, and often flushing them with water from the funnel, or taking them up and cleaning them by passing them between the tips of the arms. In *Argonauta*, the paper nautilus, the brood is cared for in a very different way. The two most dorsal arms secrete the fragile calcareous 'shell' for which this genus is best known. The numerous tiny eggs – attached to branched egg strings – are carried in this shell, which thus has no homology with the true pallial shell of molluscs. It is not a house but a perambulator!

The cephalopod male duct is single and has become exceedingly specialized for the manufacture of spermatophores (Fig. 46B). The production and transfer of these sperm packets form the most elaborate features of cephalopod reproductive biology. Each spermatophore is a narrow, torpedo-shaped tube of chitin, containing a dense mass of sperm. The male conveys bunches of them to the female, after pulling them from his own genital opening by a specially modified arm known as the *hectocotylus*. After courtship and copulation they are attached to various parts of the body of the female or may be introduced into her mantle cavity. All cephalopods – *Nautilus* included – produce some type of spermatophore, and all possess some complication – it may be very intricate or relatively simple – of one of the arms, or of the tentacle crown, with which to transfer them. No cephalopod has a penis, properly so called.

We may describe the spermatophores in *Loligo*, where they have received the most careful study.[110] Each is about 16 mm long, and a dozen or so may be produced in a day (Fig. 46). A large individual may store up to 400 at one time, and they lie in a spacious pouch called Needham's sac, which surrounds all the glandular parts of the genital duct. The rest of the duct in fact opens into Needham's sac, which itself opens by the genital aperture into the mantle cavity. About two-thirds of each spermatophore houses the viscous mass of sperm. This is surrounded by an inner tunic and an outer chitinized capsule, with the space between them tensely filled with fluid. At the narrower end the sperm mass is tipped by a small cement body, followed by a long, closely coiled spiral filament. At the very end is a chitinous cap, drawn out into a thin thread which is formed as the soft spermatophore is pulled from the secreting ducts. The terminal cap and the fluid contents are kept stretched by the spiral filament. When a bunch of spermatophores is pulled out of Needham's sac by the hectocotylus arm, the caps are loosened and the spiral filament is dislodged by the tension of the attachment threads. The filament is not – as sometimes thought – an explosive device: ejaculation of sperm is caused by the elastic contraction and osmotic action of the capsule wall. The cement body is forced out first and by this means the sperm mass which follows it is securely fixed to the female.

Fig. 46 illustrates the regions of the male duct where the parts of the spermatophore are laid down and fashioned. The two *mucilaginous glands* secrete the axial material, i.e. the mucous matrix of the sperm mass, the cement body and the filament. The inner and outer tunics come from the middle reaches of the duct. The whole structure is then thrust into a diverticulum which has been incorrectly called the 'prostate'. This serves as a hardening gland. Finally, in a terminal *finishing gland*, the cap and thread are secreted, and the spermatophore passes into the storage sac.

In *Loligo*, the hectocotylus organ, on the fourth arm on the left side, is very slight. Several of the suckers are merely modified to form an attachment area for the spermatophores. Copulation is preceded by courtship ritual. The male swims alongside the female and displays at intervals, spreading his arms and assuming a dark red blush. There are alternative positions of copulation. The male may come to lie parallel to the female, lower sides in contact, grasping her by wrapping his arms tightly round her head. With a sweep of the hectocotylus arms he plucks out a bunch of spermatophores and inserts them into the female's mantle cavity, where they are attached near the oviduct. In the second position the male and female join head to head and the spermatophores are transferred to a glandular patch on the female's buccal membrane, which may be identified as a 'receptaculum'. The sperms are still immobile. They are activated by contact with sea water when their reservoir is ruptured at the time the eggs are laid. Courtship and copulation are communal activities in the squid. Large numbers of males and females – which are at other times segregated – gather together at the spawning ground. The females contribute hundreds of egg-masses to the community pile.

As in a number of other cephalopods, oviposition appears to be rapidly followed by death.

Very interesting and diverse are the hectocotylus arms in other cephalopods. In *Nautilus* a special portion of the tentacle crown is hectocotylized, involving four tentacles of one of the lobes on the right side. This is called a *spadix*, forming a sleeved projection with a glandular tip. In some squids, such as *Rossia*, all the suckers of the hectocotylus arm are lost and are replaced by a glandular adhesive membrane. It is in the Octopoda that the hectocotylus is most elaborate. The third arm on the right is furnished with a spoon-shaped tip and this is connected with the base of the arm by a fold of skin which appears to form a sperm groove. The female is first caressed at the full arm's length by the male, and the tip is then put into her pallial cavity and the sperms deposited at the mouth of the oviduct. In Argonautacea (*Argonauta, Tremoctopus* and *Ocythoe*) the male is much smaller than the female and the hectocotylus is enlarged and autonomous. It has a long filament which remains coiled

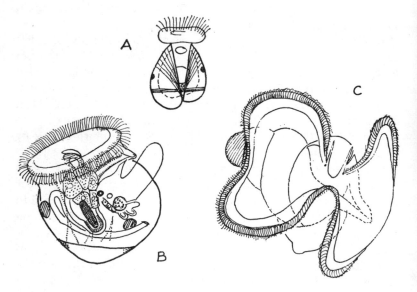

Figure 47
(A, B) Oyster larvae, in end and lateral views
(C) veliger larva of the gastropod *Philbertia* (*after Fretter*)

in a sheath until the arm is detached, after which the whole organ moves about freely for some time in the mantle cavity of the female. It was long regarded as a parasite, even by Cuvier, who gave it a special name, 'Hectocotylus octopodis'. Later and more imaginative workers endowed it with a gut, heart and reproductive system! Sex dimorphism is most extreme in *Argonauta*, where the female carries the papery shell. The male is a small dwarf one inch long, his largest organ being the detachable hectocotylus arm.

Life history and larvae

Two purposes may be served by larvae – finding new settling sites and gaining access to the rich food supply of the phytoplankton. Molluscan larvae are of different types according to the importance of the pelagic phase and the amount of planktonic food taken. The earliest larva was undoubtedly a trochophore like that of an annelid – a top-shaped creature with a tuft of cilia above and ciliated band around the middle; and this is perhaps the closest resemblance that the Mollusca have ever borne to the Annelida.

Molluscan life histories do not perfectly correspond with taxonomy, but it is in general true that the archaeogastropods and the bivalves begin life as a trochophore and rapidly pass on to a veliger. Fig. 48 shows the typical larval organs – most prominent a ciliated velum, drawn out in lobes in gastropods from the ciliated ring lying in front of the mouth. In the gastropods the early veliger goes through the critical episode of torsion, and its velum may later become very large, often subdivided in prosobranchs into four or six lobes, often ornamented by coloured spots. In bivalves a wheel-shaped velum projects from between the two valves of the larval shell. It is never deeply subdivided as in gastropods.

Thorson has recognized three ecological types of larvae, each represented in molluscs. First there are planktotrophic larvae with a long larval life of up to two or three months, as in most lamellibranchs and many prosobranchs. These are the least modified larvae of all, and are small and cheap to produce on a large scale. (*Mytilus edulis* may spawn 12 million eggs, all capable of developing into larvae.) They are effective pioneers and the species which use them secure wide distribution. They will settle in great numbers in good years, and the stock of benthic adults is thus subject to great fluctuations according to seasonal conditions. Planktotrophic larvae are

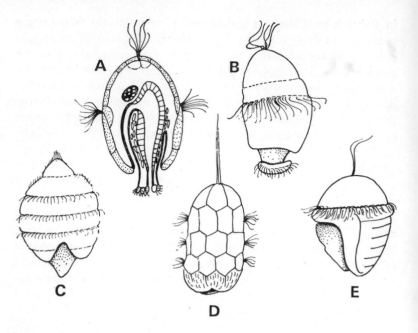

Figure 48 Some molluscan 'yolk larvae' with barrel shaped tests
(A, B) *Neomenia,* in vertical section and entire
(C) *Dentalium*
(D) *Yoldia*
(E) *Chiton*

most usual in tropical and sub-tropical seas, and in high arctic seas the few molluscan larvae found are also of this type.

Such molluscan veligers are all ciliary feeders. The large velar cilia collect particles which are thrown on to a tract at the base of the velum leading to the mouth. Coarse or unsuitable particles are removed by rejectory tracts upon the foot.

Secondly, there are planktotrophic larvae with a short swimming life of never more than a week in the plankton. The velum is never elaborate, and planktonic feeding is of secondary importance, distribution being the main object of larval life. There is little growth between hatching and settling. Nudibranch larvae may be often of this type, and other British examples include *Gibbula cineraria, Hydrobia ulvae, Turritella communis* and *Bela trevelyana.* Such larvae are seldom if ever found in lamellibranchs. Being less dependent on food they are for small larvae surprisingly adaptable to unfavourable conditions, and serve mainly for dispersal.

The third type of larvae take no food in the plankton. They are *lecithotrophic*, hatching from very yolky eggs and developing into large, rather clumsy 'yolk larvae'. They swim little and are passively carried about in the plankton. Being independent of adverse conditions, species recruited in this way show very constant numbers from year to year. The chief disadvantage of yolk larvae is their small numbers and expensiveness to produce, and the small ability of the species to seize the chances offered by a good food year. Gastropods show few examples of yolk larvae, which are the normal type in three groups, the Amphineura, Scaphopoda and protobranchiate Lamellibranchia. The yolk larvae of chitons are modified egg-shaped trochophores with a broad ciliary ring, spending only six hours to a few days in the plankton. In other yolk larvae, large flat velar cells surround the body like a girdle. *Neomenia* has three such rings, *Dentalium* four. In protobranchs such as *Yoldia* and *Nucula* they form a large barrel-shaped ciliated test, which is thrown off when the larva settles (Fig. 48D).

Not all molluscs have a planktonic life history. With the poor phytoplankton of arctic seas, few molluscs employ larvae, and yolk larvae are, surprisingly, absent. If larvae are to be liberated there at all, as in *Mya truncata* and *Saxicava arctica*, they must be numerous and planktotrophic, and the species must accept the hazards involved in taking advantage of the brief plankton bloom.

Many gastropods pass their whole development in the egg capsule. This is inescapable in land forms and is the rule in freshwater snails as well. In the sea too there are many examples of non-pelagic development. Most neogastropods (except Nassariidae and turrids) retain the larvae in the eggs, as do many mesogastropods and even some archaeogastropods. Free-swimming larvae are nearly always produced in that thoroughly marine group – the Opisthobranchia.

With some retained larvae the egg's own yolk is sufficient to carry it through development within the capsule, as in *Lacuna pallidula, Littorina obtusata* and *L. rudis* (*L. littorea* and *L. neritoïdes* are planktotrophic). But most non-pelagic prosobranchs are fed on nurse eggs which are enclosed in the same capsule but do not develop. *Natica catena* has usually fewer than 10 nurse eggs per embryo, and may sometimes produce pelagic young; *Nucella lapillus* has 20–30, *Buccinum undatum* 100, *Sipho islandicus* 7000 and the deep-sea *Volutopsis norvegica* up to 100000!

A crisis in molluscan life must come at the settlement stage, when the pelagic larva metamorphoses. It was indeed once thought that only those that first touched down on a suitable ground could survive: those which fell by the wayside had no second chance. Major work discussed by Gunnar Thorson on invertebrate larvae has shown, however, that the first choice may not be irrevocable. Larvae such as those of worms reared by D. P. Wilson and the echinoderms of Th. Mortensen were able actively to select a favourable substratum. Marie Lebour relates that the larvae of the molluscs *Nassa*, *Philbertia*, *Rissoa* and some *Natica* species passed through a short ambivalent swimming-crawling stage, when they could sample the substratum and select a site with great precision. Thorson also found that the larvae of the boring bivalve *Zirphaea* penetrated into the cork floats of collector bottles and settled there only. There is perhaps a critical trial period – not longer than a week – during which most larvae can postpone their final settlement while in search of a site. We find examples in sessile molluscs, as in oysters, of the settlement behaviour described by Knight-Jones and Crisp in barnacles and *Spirorbis*: here the larvae are strongly induced to settle by the presence of their own kind – even dead shells or persistently detectable traces of their own species previously on the site.

As we have seen, larvae are not always reliable guides to phylogeny, and may evolve many structures for their own needs as distinct from the morphology of the adult. This we call 'clandestine evolution'. The embryo shell can, however, give many clues as to both the relationships and the mode of life of its occupant. Thus the size and number of the embryonic whorls differ according to whether the larva had a long or short planktonic life. A multispiral protoconch with a small apex and many whorls denotes a long-swimming larva, and is thus characteristic of most gastropods of warm seas. A species with a yolky larva – more typical of high latitudes – has a simple protoconch with a large bulbous apex. By inspecting the protoconch of an adult (even a fossil) shell, we may thus deduce the mode of larval life. This rule, first discovered by Dall, applies even among the species of a single family. Thus the Naticidae and the Struthiolariidae both show alternative types of apex depending on different lengths of life history.

In a number of gastropods the embryonic shell is sinistral, and reverses its direction to coil to the right in the adult. Such an apex is evidently non-adaptive, and is found in both early opisthobranchs (bulloids and pyramidellids, while *Limacina* retains it as the adult shell) and in early pulmonates such as Ellobiidae and Chilinidae. It is an interesting confirmation of other evidence of the common origin and later divergence of these two groups of gastropods.

Phylogeny of larvae

From Salvini-Plawen's discussion of the relations of the Aplacophora with the earliest molluscs, it seems likely that the primitive larval type within the phylum was a trochophore that had become considerably modified into the barrel-shaped 'test-cell larva', still to be found in variant forms in the Aplacophora, Scaphopoda, chitons and prosobranch bivalves. At an early level, 'pre-adult' stages would have settled direct from the test-cell larva, but in higher Mollusca a further free-swimming stage is held to be inserted, with the expansion of a ciliated band or 'proto-troch' into the wide, broad-lobed velum anterior to the mouth.

In a detailed review of the gastropod veliger, Fretter and Montgomery have shown how the velum functions both for swimming and food-collecting. With the veliger becoming the dominant larva in the gastropods, and in a variant form in the bivalves too, the trochophore stage is suppressed or eliminated. From the minutely coiled protoconch in *Neopilina*, there is evidence that the Monoplacophora – or some of them – had also acquired a dispersive phase corresponding to the veliger.

8 Nervous system, sense organs and behaviour

In no other phylum except the Chordata does the evolution of the nervous system cover such a wide span as in the Mollusca. An early mollusc such as a chiton is a slow-moving creature with a nervous system best compared with that of a flatworm; but the highest productions of the Mollusca – the cephalopod brain and sense organs – are rivalled only among the vertebrates themselves.

Much exquisite dissection has from the earliest days been devoted to the molluscan nervous system, together with compendious illustration and description. Such work has been recently marshalled and freshly presented in the magnificent book on *The Nervous Systems of the Invertebrates* by Bullock and Horridge.[14]

In the chitons – as no doubt in the first molluscs – neuromuscular action belongs chiefly to the head-foot; the muscles of the sole and the odontophore are from the first very complex. In the visceral mass and pallial cavity the chief effector organs are cilia and mucous glands, and these parts are hardly as yet brought under rapid control. In the primitive plan of the nervous system (Fig. 49A) there are few localized ganglia. The nerve ring round the oesophagus is built up of a dorsal cerebral band with scattered neurones, and a ventral labial commissure, sending forward connectives to a pair of buccal centres. These ganglia control the movements of the odontophore, and were probably the first to become distinct. Two pairs of parallel cords run back from the nerve ring, also with scattered nerve cells, and linked together by cross commissures in ladder fashion. These cords are the foundation of the viscero-pallial and pedal nervous system respectively. The pedal cords run along the foot on the floor of the perivisceral space, and the two pallial cords lie laterally near the attachment of the mantle to the body. In addition, a simple stomatogastric or 'sympathetic' nervous system runs back from the nerve ring along the wall of the gut.

The nerve cells in the longitudinal cords soon become localized in distinct ganglia. In the earliest gastropods we find already a pair of

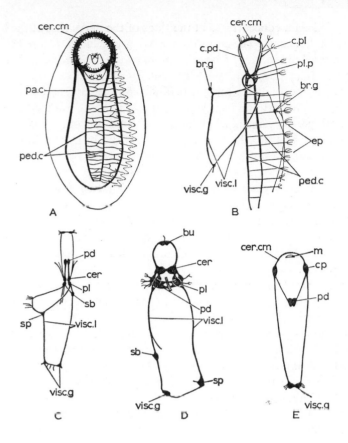

Figure 49 The nervous system in Amphineura, Gastropoda and Bivalvia
(A) A chiton
(B) *Haliotis*, an archaeogastropod
(C) *Triton*, a mesogastropod
(D) *Akera*
(E) A eulamellibranch
br.g, branchial ganglion; bu, buccal ganglion; cer, cerebral ganglion; cer.cm, cerebral commissure; c.pd, cerebropedal connective; c.pl, cerebro-pleural connective; cp, cerebropleural ganglion; ep, epipodial nerves; m, mouth; pa.c, pallial cords; pd, pedal ganglion; ped.c, pedal cords; pl, pleural ganglion; pl.p, pleuropedal ganglion mass; sb, subintestinal ganglion; sp, supraintestinal ganglion; visc.g, visceral ganglion; visc.l, visceral loop

pleural ganglia concentrated at the sides of the nerve ring at the head of the pallial cords. The cords themselves are reduced to connectives, forming a long *visceral loop* with a *parietal ganglion* becoming distinct on either side, and a single or sometimes paired *visceral ganglion* where the loop crosses the gut behind. *Haliotis* or *Trochus* may be chosen to illustrate this early condition (Fig. 49B), with no distinct cerebral ganglia and the pedal ladder still unconcentrated. The cerebral ganglia are the next to appear, while in most mesogastropods the neurones of the pedal cords have been drawn into the nerve ring as a pair of ganglia; while the cords as such disappear. Thus we find the typical gastropod nerve ring, with two cerebral ganglia dorsally and two pleural and two pedal ganglia below. The cerebral and pedal pairs are linked in the middle line by cerebral and pedal commissures; and at either side the cerebrals are linked with the pleurals and pedals, and the pleurals with the pedals, by short connectives.

From the pleural ganglia springs the visceral loop, connecting to the nerve ring the parietal ganglia, from which nerves pass to the pallial organs. During torsion the gastropod visceral loop is twisted into a figure-of-eight, the condition known as *chiastoneury*. The right parietal ganglion now crosses above the gut to lie upon the left, being called the *supraintestinal ganglion* (though it really lies near the oesophagus). The original left parietal ganglion passes underneath the oesophagus to lie on the right as the *subintestinal ganglion*. In the Monotocardia the supraintestinal ganglion takes on the innervation of the remaining left side of the mantle cavity, including the surviving gill. It also supplies the osphradium, either directly or through an outlying *osphradial ganglion*. On either side the pleural and parietal ganglion may establish a direct link by a connective known as the *zygoneury*.

In the higher Gastropoda all the ganglia finally come to lie in the nerve ring, and the figure-of-eight configuration of the visceral loop is lost. First – in the higher prosobranchs such as *Buccinum* – the subintestinal ganglion is withdrawn across to its original side to lie close against the left pleural ganglion. Then the supraintestinal ganglion moves back over the oesophagus to become attached to the right pleural member. In the pulmonates the whole visceral loop is shortened with the incorporation of the visceral ganglion in the ring, which now consists of nine large ganglia, with the smaller buccal ganglia still attached by connectives. Torsion of the visceral mass has not been abolished in the pulmonates, but the evidence for it – in the twisting of the visceral loop – can no longer be clearly detected, and in *Helix* the nerve ring is very compact, with its ganglia in direct contact.

The Opisthobranchia (Fig. 49D), however, have completely reversed torsion. During their phylogeny the mantle cavity has moved backwards along the right side towards its original site and the visceral loop actually untwists. All the ganglia lie in the nerve ring, usually above the oesophagus and united only by the pedal commissure below. The early opisthobranch *Actaeon* and the primitive pulmonate *Chilina* are interesting transitional forms in the reduction of the visceral loop: they still display the well-marked figure-of-eight.

Sense organs (Amphineura and Gastropoda)

In chitons the mantle and girdle press close to the ground. Head tentacles or eyes would have little use, though the snout possesses tactile organs, and the buccal cavity a gustatory subradular organ. Paired osphradia lie at the rear, usually near the last gill. Otocysts are lacking in chitons and Solenogastres, quite uniquely in molluscs, for a pair is present in Monoplacophora. The most important sensory area consists of tactile organs and light receptors, transferred where they are most needed, to the exposed surface of the shell plates. The valves are pitted with very numerous sense organs, like microscopic bright spots, and are of two kinds. The most abundant are simple epidermal papillae called *micraesthetes*. Moseley, who has studied them most closely, regarded these as merely touch receptors. The larger receptors, the megalaesthetes, lie at the centre of clusters of micraesthetes. These are much more complex, forming in many chitons, such as the non-British *Acanthopleura*, *Tonicia* and *Liolophura*, undoubted eyes, with a simple lens, a pigmented sheath and a retina. Their outer covering cuticle forms an elementary cornea. The megalaesthetes are innervated from the pallial cords and are densest near the growing points of the shell plates, where they have not yet been worn away. They are roughly arranged in rows, and one species, *Corephium aculeatum*, has some 3000 of them on the anterior valve, and on the other valves together about 8500. The dorsal sense organs dominate the whole behavioural pattern of chitons, and even those species which lack true eyes are highly sensitive to light and shade. The animal wanders afield and grazes at night or when the tide is in, and on exposure to strong light retreats under stones,

F

seeking a position of minimum light intensity, maximum humidity and maximum dorsal contact.[118]

With reorganization of the body, molluscan sense organs have often shifted to new locations. This is especially so of light receptors, which range from the simplest pigment spots to the intricate and beautiful eyes of heteropods, scallops and especially cephalopods. Like other sense organs the eyes have been many times lost, re-acquired and rearranged, according to adaptive needs.

A gastropod like the top-shell *Monodonta* shows all the typical sense organs of the early mollusc. The cephalic tentacles are tactile and probably gustatory, and share their functions with the epipodial tentacles which come into wide contact with the ground. At the bases of the cephalic tentacles are paired eyes, and embedded in the foot are simple otocysts with calcareous otoconia. On the inhalant side of the mantle cavity lies the distinctive pallial chemoreceptor, the osphradium.

In later Gastropoda these organs have been emphasized or reduced according to habits.[2] The eyes are primitively simple in the limpets, forming open retinal pits with neither lens nor cornea. In other archaeogastropods, such as *Trochus*, *Haliotis* and *Turbo*, the optic vesicle has only a narrow opening and is filled with a watery humour; in higher gastropods it contains a spherical lens and is closed by a double-layered epithelial cornea. The majority of prosobranchs probably use their eyes for simple orientation in light, in the same manner as was demonstrated by Fraenkel with the sacoglossan opisthobranch *Elysia viridis*. This small slug moves in a constant direction in horizontal light; the angle between the axis of the body and the line joining the eye to the source of light is known as the 'orientation angle'. Only light falling on the cup-like eye within 35° and 130° of the body can reach the retina, so that orientation angles outside these limits cannot be used. The eye thus serves as a 'light compass', a function which must be widespread, and has recently been described by Newell in the periwinkle *Littorina littorea*.[236]

The greatest visual powers are found in two groups of fast-moving gastropods. The pelagic Heteropoda have – as we have seen – tubular telescopic eyes, with a large lens, a tapetum and much-folded retinal surface. As in Cephalopoda, accommodation is effected by altering the distance of the lens from the retina. Heteropods find their direction and capture food by visual means, and the osphradium is quite vestigial. In the bottom-dwelling Strombidae, which progress with powerful leaps, the large eyes are mounted on long peduncles which dwarf the cephalic tentacles. The animal has an aspect of great alertness. Exploration is rapid and visual, no longer by the slow, tentative action of the tentacles, and the optic peduncles peep vigilantly from under the canopy of the shell, being moved actively to and fro.

In burrowing gastropods like the Naticidae, the Olividae and bulloids such as *Scaphander* and *Philine*, as well as in many nudibranchs, the eyes are small and buried in the skin, or altogether lacking. The passively drifting pelagic prosobranch *Ianthina* differs strikingly from the heteropods in being blind, and in having also lost the statocysts. In the Heteropoda the statocysts are highly developed as organs of balance. The sensory cells form a large macula and there is a single large otolith. After removal of one statocyst, *Pterotrachea* is at first unable to orient, developing a pronounced roll towards the operated side.

One or two pelagic opisthobranchs have compensated for the sacrifice of the eyes during their earlier benthic history. Thus the shelled pteropod *Corolla* has special adaptive eyes, complete with lens and retina, appearing as pigment spots round the edges of the wings. The bottom-dwelling onchidiid slug *Peronia* has followed the chitons in shifting the light receptors to the dorsal surface, which is studded with small tentacles bearing eyes remarkably like the pallial eyes of some lamellibranchs.

The land pulmonates explore much more by sight. The eyes are carried at the tips of long inversible head tentacles, and some snails are found to turn aside from an object at a distance of 10 cm. Such eyes are, however, small and simple in structure, and form perception is probably lacking, as in all molluscs except cephalopods. The tactile and chemosense has shifted in pulmonates from the large tentacles to a pair of small secondary tentacles developed from the oral lappets.

The forward-facing mantle cavity in prosobranchs becomes – rather in the manner of the protochordate pharynx – the centre of all those functions depending on the passage of a water current: respiration, olfaction, detection and removal of sediment, and in a few cases feeding as well. The mantle is highly sensitive to inborne particles, and Hulbert and Yonge have held that the osphradium, as well as being a chemoreceptor, may also be a mechanoreceptor detecting particles suspended in the water current. Herbivorous gastropods such as trochids and littorinids, even those like *Aporrhais* and *Turritella*, that dwell near silt, have nevertheless relatively simple osphradia. This organ is largest in the carnivores, becoming broad and filamentose like a subsidiary gill (p. 82). The inhalant siphon, which leads the water current to the osphradium, now becomes the anterior outpost of the body, acting as a moving nostril continually sampling the environment ahead. Copeland has shown that the whelks *Alectrion* and *Busycon* begin to crawl or accelerate in the direction of oyster juice, and find their prey by a directed reaction or klinotaxis. *Nassarius reticulatus*, according to Hentschel, has the same ability. Whelks such as *Cominella* can detect carrion from a distance of six feet, and the Conidae (p. 108) use their siphon and osphradium for stalking live prey.

With the lack of a pallial water current in opisthobranchs and pulmonates the osphradium ultimately disappears too. The typical olfactory organs of opisthobranchs are a pair of modified head tentacles, the rhinophores, which become clubbed and finely plicate to increase the sensory area. The tactile sense reaches a new importance in naked nudibranchs, the whole dorsal with its processes being highly sensitive to contact.

Both chitons and limpets show a homing behaviour, involving a wonderful power of orientation. After a feeding sortie, at night or during high tide, they return with great precision to their permanent resting site. Many limpets excavate a deep scar, its edges eroded by the shell and conforming exactly to it in outline. The distance of the possible journey varies with the species. Stephenson – see Thorpe (1956) – finds that *Patella granularis* travels and returns from as far afield as 1½m (5 ft). Accuracy of performance falls off with distance. Chitons, *Onchidium* and the marine pulmonate *Siphonaria* make similar journeys and, sometimes at least, may return by a different route. On reaching the scar or resting site the animal turns and manoeuvres till an exact fit is obtained. This happens even after the scar has been reversed in position, or when the whole rock has been lifted out and reoriented. Such behaviour implies a space memory and a subtle appreciation of topography not yet explained by the available sense organs. *Siphonaria* and *Onchidium* are said to use touch, the one with the edges of the mantle, the other with the buccal lips. Removal of the tentacles with the eyes in limpets does not prevent homing. Tactile information must play a large part, but any detailed registering of the substratum by a 'touch memory' would seem to be ruled out by the experiments of reorienting the scar, and by the different homeward path sometimes taken. The problem is complicated, and obviously a wide appreciation of many aspects of the environment is involved. To formulate this in terms of classical neurology is not easy, since a gastropod mollusc has only relatively few neurons available in its total 'brain'.

Bivalvia

The bivalve nervous system is essentially simple. Paired cerebral ganglia lie above the oesophagus and – with the exception of the protobranchs – the pleural ganglia are completely fused with them. Pedal connectives lead to the pedal ganglia embedded in the base of the foot, and long visceral connectives run back to a pair of visceral ganglia lying beneath the posterior adductor muscle. The cerebropleural ganglia innervate the palps, anterior adductor muscle and part of the mantle, as well as the otocysts and osphradia. The visceral centres control a large territory: they innervate the gills, heart, pericardium, posterior adductor muscle (sometimes the sole adductor remaining), as well as part or all of the mantle, the siphons and pallial sense organs. Where concentration of the nervous system has taken place, as in *Pecten*, *Spondylus* and *Lima*, it is the visceral ganglia that provide the new centre and the cerebrals that move back to join them, forming a 'visceral brain' of an elementary sort, sometimes with distinct optic lobes receiving nerves from the pallial eyes.

No bivalve has a head and most of the sense organs have withdrawn from the anterior end. Vestigial eyes are found here only in Mytilidae and *Avicula*. The otocysts are very simple, deeply embedded in the foot near the pedal ganglia. In protobranchs and the Mytilidae they have the primitive form of otocrypts, opening narrowly to the exterior. In the sedentary Ostreidae they lack otoliths. The paired osphradia are simple sensory patches lying near the attachment of the gill close to the visceral ganglia, though they are innervated by cerebral nerves running through the visceral connectives.

The rest of the exteroceptors have come to lie at the edge of the open mantle or – in burrowing bivalves – are concentrated at the tips of the siphons. Thus *Cardium* possesses sensory tentacles round both siphons and these are equipped with small but rather complex eyes, with a retina, hyaline lens and cornea. Few other eulamellibranchs have eyes, but the siphons commonly have pigmented light-sensitive spots, as in pholads, some venerids and in *Mya*. In sessile or surface-dwelling lamellibranchs, pallial eyes are often abundant, distributed like tactile tentacles along the free middle lobe of the mantle edge. In the Arcidae the individual eyes are of simple structure, but gathered together in faceted groups, like a compound eye.

By far the most alert of bivalves are the swimming Pectinacea. Here the pallial tentacles are highly developed, forming sensitive fringes of tactile organs or guarding tentacles for the exclusion of sediment. Most beautiful of all are the several rows of long coloured retractile tentacles in the Limidae. In *Pecten* and *Spondylus* the tentacles are shorter, many of them tipped with small but elaborate eyes, with a metallic blue sheen from the tapetum or light-reflecting layer behind the retina. The dioptric apparatus is a corneal lens, with sometimes a dome of hyaline cells lying outside it. The retina, as in vertebrates, is inverted, that is, the light must pass first through a nervous layer to the sensitive receptors beneath – an outer layer of rods and an inner layer of cones. Despite their complexity, the eyes of *Pecten* can make no satisfactory definition of shapes. They detect changing light patterns or movements, reacting to flashing lights or moving stripes at particular speeds of movement characteristic of the scallop predators, starfish and whelks. The pallial tentacles at the same time are chemosensitive; a scallop will react to starfish juice and also to speed of movement by rapid swimming (p. 203) but makes no flight response to a stationary starfish model. Horizontal swimming in *Pecten* with left side down has led to an asymmetric development of the statocysts: though both are present the orientation and control of all swimming reflexes are initiated from the left one alone.

Apart from the foot, the muscular activities of bivalves are few and stereotyped. First, there are the movements of the mantle edges and siphons, largely directed by respiratory needs. Second, and universally important, are the powerful contractions of the adductor muscles, by whose action the valves are tightly closed or allowed to gape. By these muscles, the bivalve controls its whole relations with the world outside. In most lamellibranchs, the two parts – tonic and phasic – of the adductor cannot be separately identified. But they are very distinct in those surface-dwelling Anisomyaria that retain only the posterior adductor. Here the single enlarged muscle produces – by its phasic part – the rapid opening and closing so important in expelling sediment. Sustained tetanic contractions by the 'slow muscle' will be best developed in intertidal bivalves like *Mytilus* that close tight for long spells. By contrast, the 'fast muscle' is most useful in forms like *Pecten* that expel sediment, or even swim, by the clapping of the thin, light valves, an accomplishment that has arisen from the normal adductor rhythm.

Cephalopoda

The cephalopod brain has become so highly differentiated that it is well not to look for too precise homologies with the ganglia of other molluscs. The nautiloid brain (Fig. 51D) consists of three half hoops

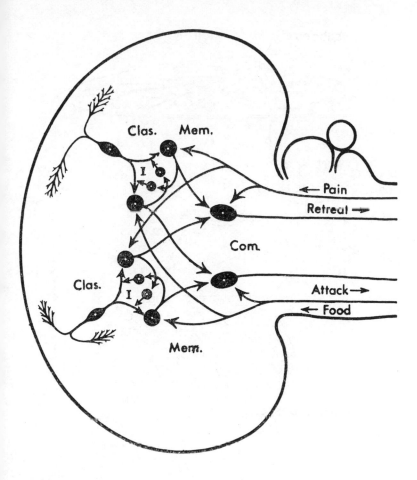

Figure 50 J. Z. Young's model of the memory system in the optic lobes of *Octopus*

Objects are analysed for shape by scanning of their horizontal and vertical extent by classifying neurones (Clas.). The information is passed to memory cells (Mem.). These – in their untrained state – would pass on to command or motor neurones (Com.) the signal to attack. If the attack succeeds in securing food, fibres signalling this result activate the memory cells, increasing the excitability of those promoting attack in that situation. This may be achieved by the activation of small inhibitory cells (I) to prevent firing of memory cells whose output is to the motor cells governing retreat. If the attack is punished, the memory cells receive a pain input; and those governing retreat are activated, and in turn activate the inhibitory cells that suppress attack

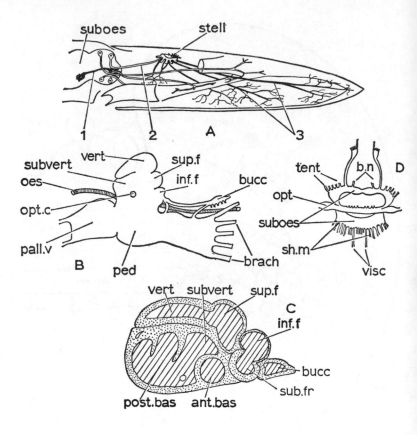

Figure 51 The cephalopod nervous system
(A) *Loligo*, giant fibre system, with 1,2,3, first, second and third order axones (*after Young*)
(B) *Sepia*, principal lobes of brain, from right side (optic lobe removed)
(C) *Octopus*, section of supraoesophageal lobes of brain (*after Wells*)
(D) *Nautilus*, brain from above (*after Owen*)
ant.bas, anterior basal lobe; b.n, buccal nerves; brach, brachial nerves; bucc, buccal lobe; inf.f, inferior frontal lobe; oes, oesophagus; opt.c, optic commissure; opt, optic lobe; pall.v, palliovisceral mass; ped, pedal mass; post.bas, posterior basal lobe; sh.m, shell musclenerves; stell, stellate ganglion; sub.fr, subfrontal lobe; subvert, subverticalis lobe; suboes, suboesophageal mass; sup.f, superior frontal lobe; tent, tentacular nerves; vert, verticalis lobe

of nervous tissue, a wide cerebral band above the oesophagus, connecting with paired optic lobes, and two bands below, an anterior pedal and a posterior pleurovisceral (see p. 178). The whole brain lies within a widely fenestrated cartilage capsule.

In later cephalopods the brain is much more concentrated and more completely enclosed by cartilage. The diagram (Fig. 51c) of the brain of *Octopus* will serve to identify the chief parts.[352] There is first an upper supraoesophageal region, divided up into eight separate centres, certain of which receive sensory nerves, as well as connectives from the buccal ganglia. The dorsal region connects at the sides with very large stalked optic lobes, which lie at the bases of the eyes and constitute a separate part of the brain. The lower supraoesophageal region of the brain consists of a pedal centre in front and a pleurovisceral centre behind, continuous with each other and linked with the supraoesophageal brain above. The pedal centre sends eight large nerves to the arms (in decapods ten nerves arise from a more or less separate branchial ganglion) and also supplies the funnel. From the pleurovisceral centre run back the two great pallial nerves to the stellate ganglia (Fig. 51A) on the inside of the mantle. These ganglia innervate the circular and longitudinal pallial muscles that contract to expel water in swimming. Two large visceral nerves also run back to the visceral mass.

Considering the motor functions of the cephalopod brain first, we shall examine the pallial innervation in a squid like *Loligo,* a much faster swimmer than *Octopus* and with a rather less modified nervous system. In *Loligo* (Fig. 51A) the pleurovisceral centre has a median ventral bulge, the giant fibre lobe, and it is from here that the contractions of the mantle are ultimately controlled. Impulses are set up in one or other of a pair of giant neurones situated in this lobe, and studded with some hundreds of cell bodies. Other large motor neurones have their cell bodies here and motor axones proceed from them to the arms, especially the long tentacles which make lightning capture of the prey. Of the two largest neurones in the giant fibre lobe, known as *first order giant neurones,* each sends a stout axone to the main palliovisceral centre, and these, after making a fusion in the mid-line (a very unusual condition in neurones), make synapses with several *second order giant neurones.* These have cell bodies in the palliovisceral centre, and send axones to the funnel and by way of the pallial nerves to the stellate ganglia. In these ganglia lie the cell bodies of *third order giant neurones,* whose axones in turn run directly to the pallial muscles.[344]

Such a giant fibre system is an ideal one for producing rapid movement rather than the prolonged contractions of tonus. There is little fine gradation of action, but excitation is synchronized on both sides and a large number of muscle fibres contract simultaneously, reaction time being greatly accelerated. Giant fibres are hence found in, for example, annelids, nemertines, crustaceans, as also in molluscs, in situations where quick escape or attack is required. In *Loligo* the mantle contractions are mediated from the brain with the fewest possible units, namely one pair of giant fibres, and with a rapid conduction rate, attaining up to 20 m/sec. One combination not achieved by cephalopods is that of speed with gradation or precision of movement; this is possible only in vertebrates, with rapid conduction by numerous nerves of normal size, by virtue of the special properties of the myelin sheath.

In the slow-moving *Octopus* there are no giant fibres to the stellate ganglion, and impulses travel to the mantle by axones of ordinary size. In addition some of the neurones within the stellate ganglion produce a series of neuro-secretory processes, ending blindly in a glandular *epistellar body* attached to the ganglion. If this gland is removed locomotion is impaired, the body becomes limp and the chromatophore muscles lose tonus as well. Direct nervous control of muscle tone seems in part replaced by neuro-secretion.

There are four areas in the suboesophageal brain whose impulses produce expansion of the chromatophores. These – as we have seen – are microscopic bags of pigment, expanded by extrinsic muscles under nervous control, and contracting by their own elasticity. Two anterior chromatophore lobes – for the head and arms – are found in the pedal mass, two posterior lobes for the mantle lie in the visceropedal mass. The bottom-dwelling *Sepia* is outstanding in its speed of colour change, and *Octopus* has also a well-developed chromatophore effector system. In pelagic *Argonauta* and *Loligo* these lobes are smaller than in *Sepia* and *Octopus*, and the fibres of their neuropil seem less regularly arranged.

All these effector systems – pallial, brachial and chromatophoral – are located in the suboesophageal part of the brain, collectively comprising the *lower motor centres*. They are, however, connected with and closely supervised by the higher centres of the brain, and these we must now examine. Lying immediately over the oesophagus, and so forming the lower part of the supraoesophageal region, are the areas known together as the *higher motor centres*. There are three lobes: the *lobus anterior basalis* supervising the movements of the head and arms, the *lobus posterior basalis*, concerned with mantle and funnel movements, and the *lobus lateralis basalis*, supervising the expansion of the chromatophores.

Above the higher motor centres and lying thus at the top of the brain are five centres which are functionally as well as topographically the highest in the brain. Of these the *lobus inferior frontalis* is the collecting centre for tactile information, especially from the arms, and is larger in *Octopus* than in decapods. The two olfactory lobes are small centres in cephalopods, lying upon the optic stalks. Above all the others lie the *lobus superior frontalis* and the *lobus verticalis* with the *subverticalis* beneath it, which receive no direct sensory nerves.

The dominance of the higher centres is less obvious in *Sepia* than in *Octopus*. In a cuttlefish with the whole of its supraoesophageal brain removed, the lower motor centres can still produce a sustained forward ripple of the fins, driving the animal backwards, for as long as three days. This movement is – in itself – independent of reflex stimulation, though it can be stopped by reflexes, as set up for example by gentle touch. It appears to originate from a rhythmic flow of impulses from the lower motor centres: a simple innate behaviour is thus initiated from the brain itself – influences from sensory inflow from the world outside may modify or inhibit it, but do not promote it. It is the role of the supraoesophageal brain to regulate such elementary behaviour, chiefly by suppressing parts of it. Thus, in *Sepia*, if the lobus anterior basalis of one side is alone removed, a spinning movement results, with the operated side outwards. The impulses promoting fin movement on this side are now free of inhibition by the higher motor centres.

In *Octopus* the lower brain is much less autonomous. The supraoesophageal brain and the optic lobes together exert a close control over behaviour. If both these regions are removed, the animal becomes limp and loses all definite posture, performing only simple acts such as breathing and proprioceptive reflexes. It behaves in some ways like a 'spinal vertebrate'. If the supraoesophageal centres are so removed that the optic lobes are left connected with the lower brain, normal posture is maintained, but walking or food capture is impossible; the picture is one of 'decerebrate rigidity'. If both optic lobes and one half of the supraoesophageal brain are removed, the octopus when prodded walks in circles, the side with the intact supraoesophageal brain being on the outside. Removal of both optic lobes with the supraoesophageal brain left intact prevents any walking: the two halves of the supraoesophageal brain appear to inhibit each other. The flow of impulses from the optic lobes evidently determines the balance of activity here. If the optic lobe stalk on the left, for example, is cut, and the nerves between the retina and the optic lobe are severed on the right, the animal circles with the left side outwards. The intact right optic lobe inhibits the supraoesophageal brain on its own side. If the retinal nerves are intact on both sides, the animal never circles. Boycott and Young suggest that the retina may step up a continuous discharge of impulses that act to inhibit the inhibitory effect of the optic lobe.

Removal of the five highest supraoesophageal lobes does not seem to interfere with normal movements. Only if the three basal lobes, or higher motor centres, are damaged is behaviour abnormal. Young compares these three areas to the tegmentum of the mammalian midbrain: they are distinct from and functionally intermediate between the lower motor centres and the higher or associative centres. By the basal lobes innate or rhythmic behaviour is integrated into a total behaviour pattern. Visual information, for example, may be received from the retina, and the resulting impulses pass from the optic lobes to the higher motor centres. These centres, by their control of the lower brain, regulate and co-ordinate the elements of innate behaviour, in the light of the sensory appraisal of the animal's present environment.

In nearly all cephalopods the recognition of food or enemies depends primarily on sight. *Nautilus* alone retains osphradia, while all cephalopods have a simple olfactory pit beneath the eye. But in Decapoda the chemosenses appear poorly developed, and – in cuttlefish and squid – vision seems all-important. It is possible because their behaviour has become so highly geared to the light sense that these cephalopods must produce their own bioluminescence as a condition of life in abyssal waters. In abyssal octopods, on the other hand, there are few luminescent species: chromatophores are lacking and in *Cirrothauma* even the eyes are reduced. Where the Decapoda keep the two long tentacles ensheathed except in food-catching, the Octopoda have their eight long tentacles in constant action. Important as vision still is, the benthic octopods keep up a tactile contact with the bottom. Not only do the suckered arms perform complex manipulations, as in building elaborate retreats or 'villas' out of stones; the suckers are also extremely sensitive chemotactile organs. Blinded octopuses have high mechanotactile abilities, but even more sensitive is the power of 'taste by touch' where objects such as shellfish and test materials can be recognized by chemical differences down to 1/100 or 1/1000 the human perceptive threshold.

The statocyst of *Octopus* is comparable in complexity to the vertebrate inner ear. Its sac contains a gravity receptor formed by a macula loaded with a calcareous otolith. Angular acceleration is registered by three long cristae lying in three planes at right angles, with granule-loaded sensory hairs.[349]

If the superior frontalis and verticalis lobes of *Octopus* are removed, much of the behaviour is left unimpaired. Activities such as the reflex behaviour of mating and the associative behaviour involved in food capture still continue. But the animal will no longer search actively for food and will hunt objects only within its immediate vision. The frontalis and verticalis provide, in the words of Young and Boycott, 'a system of action wider than that dictated by the immediate environment'. Here the animal stores experience of the past, that can be called upon to modify present behaviour. These lobes are in fact 'learning centres'.

Two great bodies of experimental work have been carried out on the learning of octopods, that of Ten Cate, Young and Boycott, primarily concerned with learning by sight, and that of Wells and certain others, dealing chiefly with chemotactile learning. Visual training was first effected by contriving alternative situations involving reward for success, while discrimination was sharpened by adding a punishment for non-success. Octopuses were shown a crab alone, and a crab along with a small white disc. The crab presented alone was allowed to be eaten, and when the crab with the disc was seized a moderate punishment of a mild 6–12 volt electric shock was administered. After a period of training experimental animals can discriminate between the members of a single pair of objects, one of which is shown with the food and will soon be taken even if food no longer appears with it.

Progress in training is shown in Fig. 52, with discrimination between crabs shown alone and crabs with a 5 cm white square. Similarly shaped objects of different size can also be discriminated, for example, squares of 5 cm and 10 cm, and 4 cm and 8 cm. This is, moreover, possible irrespective of distance, even though one eye only, and not binocular vision, is used in attack. Discrimination can be learned between figures of different shapes even when they are of the same area or have the same length of outline. The experiments by Sutherland on shape discrimination[285] show that octopuses can distinguish between horizontal and vertical rectangles, but not between

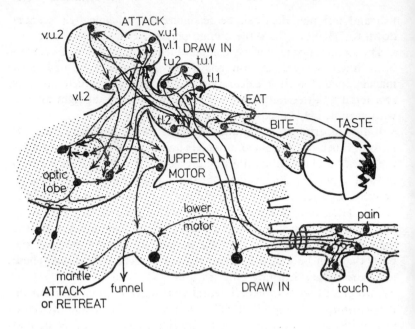

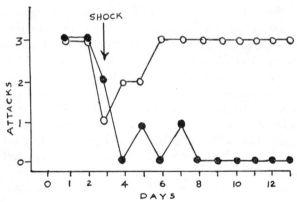

Figure 52
(*above*) Diagram of touch and visual centres, and their connections in *Octopus* (from J. Z. Young, *The Memory System of the Brain*). Receptor signalling results deliver messages to the appropriate addresses of the 'higher' brain. v.l.1, v.l.2, lower visual centres (lateral superior frontal and subvertical) v.u.1, v.u.2, upper visual centres (median superior frontal and vertical). t.l.1, t.12 and lower tacile centres (lateral inferior frontal and posterior buccal). t.u.1, t.u.2, upper tactile centres (median inferior frontal and subfrontal).

two shown obliquely at right angles to each other. There are distinct responses also to crosses and squares and – less accurately – to figures with internal differences, thus ▨ and ☐. Discrimination of squares from circles is poor. That the relation of the shape of the object to the rest of the background, or to gravity, may be of importance is suggested by the ability to distinguish ▬ from ▌ or ▨ from ◆. Such discrimination depends – it was found – on the constant orientation of the figure to the retina, so that the shape can be compared, as it were, to a horizontal and vertical grid. The pupil of the eye is in its natural position kept horizontal, regardless of the orientation of the whole animal. But injury to the statocysts upsets the orientation of the eyes and, after this operation, pre-trained animals make mistakes as if the pupils were still correctly oriented, though they may vary according to the posture in which the animal is sitting.

The learned responses of *Octopus* can be reversed after some counter-training though performance is always poorer after reversal than before. Retention is good, with maximum duration of visual memory of the order of weeks. Temporary anaesthesia with urethane does not abolish a learned inhibition. On removal of the lobus verticalis, or severing it from the frontalis superior, the inhibition was, however, lost, and negative objects were again freely attacked. Inhibition appears to depend on these two lobes or the interaction between them. Removal of the other higher centres did not affect visual learning, nor did the loss of one half of the verticalis. Moreover, operated animals had some power to re-learn and to store experience in some part of the brain left intact.

Octopus is a good animal for study in having two anatomically and spatially distinct memory stores: one visual, and the other chemotactile, from the suckers. In each, the actual memory store is accompanied by four auxiliary lobes, concerned with 'reading in' and 'reading out', sending signals to the right addresses. Signals of taste and

---

The pathway from v.u.2 to t.l.1, is the route by which the vertical lobe influences touch learning, the only connection between the two centres (*below*) Octopus trained to discriminate between a crab shown alone (○) and one shown with a 5 cm white square (●). After the first two days, a small electric shock was given each time crab + square was attacked. Three trials of each sort were made per day (from *Boycott and Young*)

pain are relayed to the memory cells. Fibres from lips and mouth interweave with those from lips and suckers. Such signals have a series of functions: first, to operate the consummatory reactions, to eat or withdraw; second to adjust the level of exploratory action, evincing caution when familiar; third, teaching the memory specifically that the results of attacking a particular object are good or bad. The auxiliary lobes spread these signals of results, and intermix them with the receptors, to reach the appropriate parts of the classifying and memory system.

In *Octopus*, two sets of four lobes are connected with the visual and tactile centres respectively. Their structure resembles the higher regions of the brain, with many minute amacrine cells without long axons.

Visual information from the eye to the optic lobe may cause an arm to be put out by activating the higher motor centres (basal lobes). Signals of taste and pain reach the optic lobe, but this system cannot by itself produce a full attack. This requires the lower visual circuit (v.l.1 and v.l.2). Taste and pain fibres also enter these lobes, and increase or decrease the possibility of attack. The upper visual circuit allows further mixing of visual signals with those of taste, increasing the tendency to attack unless pain intervenes. The upper tier of lobes (v.u.1, v.u.2) are the most intricately organized. Optic and taste fibres proceed here through the lateral superior frontal to a complex interweaving of fibres in the median superior frontal. There is more neuropil here than anywhere else in the brain. If the verticalis lobe is removed, there is no capacity to learn to attack one figure but not another if these are shown successively. The record written in the optic lobes cannot without the verticalis be properly used or read-out.

Octopus has a chemotactile learning system patterning information derived from the suckers, conveyed along axial nerve cords that bear large ganglia along their length.

The suckers of the arms in *Octopus* carry a great quantity of chemical and mechanotactile receptors.[304] In experiments on chemotactile learning, test animals were blinded under urethane anaesthesia. They would then pick up any small unfamiliar object and pass it under the interarm web to the mouth. If inedible, it is rejected, and this response is learned, a second object being taken only if sufficiently unlike the first. Rewarding with food and punishing with electric shock can be associated with objects so as to accelerate the learning process and deepen discrimination. Discrimination is made by physical texture between such objects as grooved perspex cylinders which have different proportions of grooved and flat surface. Being alike in weight and taste, these test objects are evidently distinguished solely by the proportion of the sense organs excited in the contact area, i.e. by the non-grooved part of the surface. There is no evidence of spatial projection or integration of tactile with proprioceptive information, which is used elsewhere in *Octopus*. It is thus impossible to teach discrimination of cylinders of different size, nor are shapes well discriminated though spheres and cubes are distinguished by the irregularity of the corner. Nor can weight discriminations be performed.[309]

It is unlikely that *Octopus* can ever learn skilled manipulations or improve them by practice. This is because – with the lack of any standard shape of the body – the receptors in the muscles cannot take account of all the bends and twists in the arms or the mobility of the suckers. Unlike a vertebrate or an arthropod, a labile mollusc can hardly be put into a proprioceptive strait-jacket.

The lack of reliance on proprioceptive advice can be seen from Wells' experiments with *Octopus* in mazes. Being given a glimpse of a crab through a transparent wall, the animal must then travel along an opaque walled passage out of sight of its goal, after which it will unerringly make a detour to the correct side to bring it back to the crab. Statocyst removal does not impair this ability; but unilateral blinding leads to detours on the blinded side. Bodily position is not taken into account, but the animal works by keeping close visual contact with the wall separating it from the prey.

The chemotactile centre has four lobes. No centre corresponds exactly to the optic lobes, but touch learning is no longer possible if the tactile lower second lobe (posterior buccalis) is removed. This region is very large, and the small cells seem to function as the memory store itself. The system is less specialized than the visual, and the posterior buccalis seems to correspond both to the subverticalis and the optic in the visual system. Touch learning may be a relatively recent acquisition, with no centre of its own corresponding to the optic lobes. As in the visual system, the upper circuits of the tactile system are not the memory store itself, but are needed for its proper use in difficult discriminations. It is also concerned with transfer: with the upper tactile system removed, discriminations learned by one arm or on one side are not performed by other arms.

Several rare or little known cephalopod brains have been recently described by J. Z. Young. *Vampyroteuthis* is evidently primitive, with a strange mixture of both decapod and octopod characters. The cirrate octopods may have diverged soon after this point, for they retain many primitive features, including the lack of an optic chiasma, paralleled only in *Nautilus*. Surprisingly they have giant fibres.

In *Nautilus*, studied in detail by J. Z. Young, the supraoesophageal, or cerebral cord, has nothing of the complex special centres of modern cephalopods. But the major channels are already present, vaguely discernible as if still only partially differentiated. They constitute a 'general sketch' of the finished product found in coleoids. The beginnings of learning centres exist, but no centre containing the vast numbers of minute cells found in the vertical and subfrontal lobes of *Octopus*.

The cords have no differentiation into distinct lobes. The olfactory centre is very large and the rhinophores well developed. With its pinhole eye incapable of form vision, *Nautilus* is still a macrosmatic or smell-dependent cephalopod. The gravity organ still opens by a duct to the outside, and there is no distinct crista nor macula, only a simple otolith granule.

9   The evolution of the Gastropoda

The Gastropoda are by far the largest class of the Mollusca. Modern classification breaks them into three primary divisions, the subclasses Prosobranchia, Opisthobranchia and Pulmonata. The Prosobranchia are the most numerous, widely distributed and diverse; the Opisthobranchia reach the highest level of specialization; and the Pulmonata – while more conservative in structure – show the most ambitious physiological adaptations.

The 230 families and 1640 genera* of Gastropoda form a material too rich for any tidy arrangement. Few main evolutionary lines with side-branches can be picked out. The picture is rather one of bushes branching from the base; with new shoots springing from each subsidiary stock, as the appearance of new characters provides pre-adaptations for still further evolutionary pathways. The classification of the Mollusca given on pp. 232–6 largely follows Thiele (1931),[11] with modifications introduced from Odhner's arrangement of the opisthobranchs.[240]

In reviewing the various systems in earlier chapters we have had a good deal to say about prosobranch evolution. We shall devote this chapter to certain aspects of the Gastropoda not yet treated, namely land and freshwater evolution, the radiations of the opisthobranchs, and the evolution of parasitism.

Land and freshwater Gastropoda

The first requirement of a land mollusc is a mantle cavity that can be turned into a lung, and the true Pulmonata are by no means the only Gastropoda that have left the water. The prosobranchs themselves have produced many lines of land operculate snails, and they must indeed have given rise to the pulmonates as well. To become terrestrial a gastropod must also have possessed internal fertilization and devices

*According to the highly conservative list of Thiele (1932).[11]

for storing the sperms and secreting egg capsules. This rules out most of the Archaeogastropoda, but as soon as these requirements were met, first in the Neritacea, land evolution became very persistent. Right through the Mesogastropoda, at least in those that had a reasonably unspecialized microphagous diet, every major group has had its terrestrial offshoots. Once on land, the diet could vary, even becoming carnivorous. The shell could be lost, and the respiratory arrangements altered. But the forms that made the original crossing to land were those that had not already become too committed to any specialized marine habitat.[129, 134]

The most primitive true pulmonates we know are the South American family Chilinidae and the very widespread Ellobiidae, represented in Britain by *Ovatella*, *Leucophytia* and *Carychium*. *Ovatella myosotis* lives in salt marshes, *Carychium minimum* on land and *Leucophytia bidentata* is secondarily intertidal. The Ellobiidae are not found fossil till the Jurassic, but preservation of land pulmonate remains was probably hazardous, and many technical points link this family with the prosobranchs and the earliest opisthobranchs. We must think of the Opisthobranchia and Pulmonata – widely different as they were soon to become – as having struck out originally from very similar ancestors, probably before the Carboniferous.

The headquarters of modern Ellobiidae are salt marshes and estuarine mudflats, and the largest species live in the tropical Indo-Pacific. Waters of muddy estuaries are often turbid and poor in oxygen, and it may have been much easier for molluscs to breathe air with a lung by rising to the surface than to obtain oxygen from the water by a gill. The original lung was probably a preadaptation in aquatic ancestors that made later land life possible. Other early air-breathing groups such as the Dipnoi and the Amphibia may have acquired their lungs in similar habitats in the same way. The ellobiids themselves are rather unprogressive. There are only three fully terrestrial genera: *Pythia*, which lives in tropical forests in Australia and Malaya, and *Carychium* and *Zospeum* in the Northern Hemisphere. *Carychium minimum* is very minute, and a member of the litter fauna of beech forests and other damp places. It can survive submersion and probably always lives in a saturated atmosphere, tied – like terrestrial copepods and ostracods – to a narrow sub-aquatic micro-climate. The related genus *Zospeum* is very little known: *Z. spelaeum* is blind, and burrows in limestone soil in the Karst District of Yugoslavia.

Most terrestrial molluscs have early relationships with estuarine or freshwater species, and they may share similar adaptations, especially in respiration and reproduction. Carter has suggested that freshwater life may lead to preadaptations making the transition to the land easier than by a direct path from the sea. Thus, lack of oxygen in estuarine waters may first lead to aerial respiration. Many tropical prosobranchs living in stagnant inland waters, such as *Pila* and *Ampullarius* (= *Pomacea*), have the mantle cavity partitioned into both a gill chamber and a lung. Air-breathing in aquatic snails in turn allows aestivation in response to occasional drought. This leads to further adaptations against desiccation, and a fully amphibious habit develops. With this come changes in the mode of excretion (p. 129), leading finally to a complete terrestrial life. On the other hand, cases of direct evolution from the sea to the land are rather few. A near approach is the sub-genus of high tidal periwinkles, *Melarapha*. The British *Littorina* (*Melarapha*) *neritoides* attaches by dry mucus to sun-baked rocks above high-water spring tide. It grazes on sparse algae and lichens when the rock is wet and is very responsive to splash. After five months' drought it can emerge from the shell in two or three minutes on return to water. *Melarapha* is, however, tied to the shore by a free-swimming veliger stage, and this line of evolution seems to have led no further.

A flourishing group of land and freshwater prosobranchs is the Neritacea, the highest of the Archaeogastropoda and the only ones with a special genital duct. *Nerita* itself is an amphibious high- to mid-tidal genus. On tropical coral shores its numerous species have radiated profusely to fill many niches on rocky and gravelly substrata. *Theodoxus* (= *Neritina*) lives in gently running fresh water. It has given rise to a series of snails living in fast streams where the shell has become progressively flattened and elongate: in *Navicella*, for example, we have a freshwater slipper limpet, retaining a functionless operculum on the upper surface of the foot. On the other hand, *Neritodryas*, of the East Indies, is a neritinid that has become almost terrestrial.

The next family, the Helicinidae, are entirely terrestrial Neritacea, and are equipped with a pallial lung. There are twenty-six genera, twelve of which live in the West Indies and South and Central America. The East Indies and Malaya are a second headquarters of this family. The snails live in damp places, in tropical forests, in leaf-sheaths, at the bases of epiphytes and in ground litter, and have become very diversified. *Proserpina*, evolving from *Helicina*, has lost the operculum altogether, developing like many pulmonates a series of ridges and lamellae guarding the aperture. Its shell is smooth and polished, partly enveloped by the mantle. And to underline the truth that evolution to the land is seldom one-way, we find in the genus *Smaragdia* a terrestrial neritinid that has gone back to marine life.

The land operculate family Cyclophoridae with sixty-seven genera has arisen from the most primitive stock of the Mesogastropoda, the Architaenioglossa. They are rather small, trochoid-conical or discoidal snails, living in tropical and subtropical forests like the land Neritacea. The shells of some of these terrestial genera (*Opisthostoma* and the tubed land operculates) may evolve in very characteristic ways. The freshwater counterparts of these snails are large and well known. They include the amphibious apple-snails, *Ampullarius*, of South America, and *Pila* of India and S.E. Asia. A related family is the Viviparidae, represented in slow-running muddy streams in Britain by two species. *Viviparus viviparus* is a ctenidial ciliary feeder.[90]

From near the periwinkles (Littorinacea) has arisen a further series of land snails. The Pomatiasidae, with twenty-six genera, has geographical headquarters in tropical forests of the E. and W. Indies, but has one British representative, *Pomatias elegans*, burrowing in calcareous soils. Our only other land operculate belongs to a second family, the Acmidae; *Acme fusca* is minute and long-spired like *Carychium* and lives under logs and litter in beech woods.

The Valvatidae are a small and isolated family, derived at about the same level, and living in slowly running fresh water. We have three British species with small conical to trochoid shells. Uniquely among Mesogastropoda, *Valvata* has a bipectinate ctenidium, projecting freely from the mantle cavity.[87]

Of the next superfamily, the Rissoacea, the Rissoidae are marine, with numerous genera and species, some of them having a small crystalline style and browsing on algae in tidal pools. They are even more numerous in offshore dredgings. On the upper shore, and in salt swamps, occurs a small rissoid relative, *Assiminea*; while in fresh waters in almost every part of the world occur members of the Hydrobiidae, a most numerous family with fifty-eight genera. Most are completely aquatic and like the British *Hydrobia* and *Bithynia* possess a gill.[190] A few, such as *Geomelania* in the West Indies, are land-dwellers in tropical forests.

Land operculates – according to Winkworth – account for some 4000 species of snails, as compared with more than 15000 for land pulmonates. Operculates are poorly represented in Europe, most of Asia, N. America and Africa. They come into their own in Central America, the Antilles, the West Indies, and also S.E. Asia and the East Indies. In the island of Jamaica they in fact outstrip the pulmonates in numbers. Russell Hunter gives for Jamaica the approximate figures:

Land Operculates (species)		Pulmonates
Helicinidae	120	
Cyclophoridae	30–36	approx. 215 spp.
Pomatiasidae	60	
Hydrobiidae	20–25	
Total	230–241	

The superfamily Cerithiacea is represented in Britain by such thoroughly marine genera as *Cerithiopsis* and *Turritella*, all with elongate shells. In subtropical shores many species of *Cerithium* live on estuarine mud between tides, and the family Potamididae is as a whole amphibious, browsing and trailing long shells on delta muds in Malaya and N. Australia. *Telescopium, Pyrazus, Terebralia* and some species of *Potamides* are – like the Ellobiidae – almost terrestrial and replace the gill by a pallial lung. The freshwater Cerithiacea belong to the large family Melaniidae, with thirty-nine genera (none British), which show a most interesting evolution. They are richest and most spectacular in Lake Tanganyika, an inland sea where low selection pressure has made possible a fauna of eighty-four species of snails. Of these, seventy-two are prosobranchs and sixty-six are endemic; the Melaniidae include fifty-eight of them. Their shells are variously shaped and ornamented, resembling marine genera in solidity and sculpture. Gunther, and later Moore, regarded these Tanganyika snails as a relic of a halolimnic fauna, originally marine and surviving in a cut-off arm of the Jurassic Indian Ocean. Pelseneer clearly disproved this, and showed them to form an adaptive radiation of a single melaniid stock, and Yonge has later reviewed their adaptations. Though so varied in shell form, all species are herbivores or deposit feeders with a crystalline style. Some genera, such as *Typhobia, Bathanalia* and *Bythoceras*, live at a depth of 180 m (100 fathoms) on mud. *Typhobia* bears long spines like a muricid. *Tanganyicia* and *Nassopsis* are littoral genera, living on rocks where the surf breaks. *Spekia* and *Tanganyicia* are smooth and naticoid, *Chytra* and

Limnotrochus are of trochoid shape, and *Paramelania* and *Nassopsis* are spindle-shaped, like a marine nassid.

Land prosobranchs live in a favoured environment of high temperature and high humidity. Outside these conditions operculates are not found very widely. The true Pulmonata, however, have had a more spectacular evolution and enjoy a much fuller terrestrial life. After the arthropods they are perhaps the most widely speciating and successful of land invertebrates. If pulmonates lack the structural variety shown by opisthobranchs, this is because they have made a much more subtle use of physiological adaptations. The Helicacea and the Limacidae are at once the most familiar pulmonates in this country, and the most specialized and progressive. The Helicacea are generally accounted the 'highest' group of pulmonates. The distinguishing feature of terrestrial pulmonates (order Stylommatophora) is the mounting of the eyes at the tips of inversible tentacles. In the aquatic order Basommatophora, on the other hand, they lie at the tentacle base as in prosobranchs. The family Ellobiidae are classed at the foot of the Basommatophora, and from snails at least comparable with these it is likely that both orders of pulmonate arose. Land pulmonates belonging to the primitive Endodontidae are, however, known from the Carboniferous, long before ellobiid fossils first appear. The order Stylommatophora is a huge assemblage, running to fourteen superfamilies and some 600 genera. The order Basommatophora is much smaller, with forty-eight genera grouped into four superfamilies.

In land evolution the water problem was probably the first and greatest. The smallest and earliest Stylommatophora are restricted like operculates to habitats where the atmosphere is moist. Most Endodontidae and Zonitidae, for example, live in leaf mould or litter, under dead bark, or in dark and damp places. The Succineidae are permanently amphibious and hardly able to resist desiccation. Many pulmonates, however, can tolerate intermittent dry seasons. British species of *Helicella* are found in summer, sealed by a dry mucous film and hanging from grass blades in the heat of the sun. Such African species as *Helix lactea* and *H. desertorum* congregate in thousands on dry scrub, in a mid-day temperature of 43°C. With rain they become active and creep out in marauding swarms. By burrowing in the soil where the temperature a few inches deep is many degrees cooler, various snails are able to aestivate in the dry season. The heartbeat is reduced and respiration greatly slowed down. Some species can enter a prolonged state of diapause or suspended activity when conditions are unfavourable. A famous example is a specimen of *H. desertorum* in the British Museum which after four years fixed to a tablet emerged and crawled about when taken into damp air. Comfort cites a six-year diapause in *Buliminus pallidior*, and even a claim of twenty-three years for *Oxystyla capax*.[89]

Hibernation, aestivation and tolerance of dehydration are achievements of the higher Stylommatophora. *Helix, Arion* and *Limax*, for example, were found by Howes and Wells to show a regular hydration and dehydration cycle, marked by large and irregular fluctuations in weight, as in internal osmotic pressure. In dry weather they lose water and tend to aestivate when weight is low. Aestivation is thus tied up with the natural water cycle and can also be induced experimentally by desiccation. Feeding and digestion are confined to the weight peaks and are incompatible with aestivation. The immediate mechanism of reactivation is not hydration, but a sensory stimulation effect: 'the raindrops knock at the door and the snail comes out. Hydration follows after, when it has eaten and drunk' (Wells). Unlike most terrestrial animals, slugs have no structural protection against desiccation and show a probably unique range of fluctuation in water content. Thus *Limax variegatus* was found to lose by evaporation 2·4% of its initial weight per hour (58% in twenty-four hours) while still; when actively crawling it lost by mucus secretion and evaporation 16% in one hour! This is normally replenished by feeding and drinking. Inactive *Helix* conserves water by a secreted epiphragm, and can physiologically reduce water loss from the mantle edge (see Machin).[199]

Arion and *Limax* have a diurnal cycle: their activity reaches a maximum at night, being stimulated, below 21°C, by falling temperatures. Above this point activity rises with temperature, enabling a more rapid moving away from harmful or lethal temperatures.[97]

Uricotely (p. 128) is another adaptation to conserve water. Uric acid can also be stored in the eggs, though this is little evidence that snail eggs are truly impermeable or 'cleidoic' as in reptiles and birds. They may lose by evaporation 40% of their water, a fowl's egg only 15%. Some snails however develop normally after a loss of as much as 85% (Needham). Safety is secured partly by indifference to desiccation, partly by behavioural adaptations such as burying the eggs or laying them in the shade.

The sea is never lacking in calcium, but many land habitats may be so. Land gastropods have a calcium store in the digestive gland upon which they draw for shell-building. Their shells are never as massive as marine ones, and in most pulmonate groups we find evolutionary series leading to complete loss of the shell. Thus – in the Zonitidae – we have a progressive lightening of the shell in the series *Zonitoides, Retinella, Oxychilus* and *Vitrea*. In the neighbouring family Vitrinidae the shell is extremely thin and fragile, sometimes overgrown by the mantle, and evolution leads on to the slugs of the family Limacidae (*Limax, Agriolimax* and *Milax*). By a parallel trend, the slugs of the Arionidae are derived from primitive endodontid pulmonates. Tolerance of water loss and economy in calcium are both preadaptations to the slug habit, and we have seen as well (p. 49) the mechanical advantages of the tapered and compressible body.

Of his ninety-five British Stylommatophora, Boycott lists twenty snails as obligate calciphiles, and a further sixteen as strongly preferring calcareous soils. Only one species – *Zonitoides excavatus* – is an obligate calcifuge, restricted to acid heaths and woodlands. In the fifty-eight indifferent species are included all the slugs and only one snail with a substantial shell. In analysing habits, twelve species were found to be obligate hygrophiles, thirteen to be xerophiles, eleven – including eight slugs – anthropophiles near human cultivation, the remainder preferring woodlands.

There are, however, few land habitats where some land pulmonates have not spread. There are numerous desert xerophiles, and several burrowing forms, for example, in Britain, *Caecilioides acicula* and the testacellid slugs. Like the land operculates, the Pulmonata are most diverse and of largest size in the hot, moist forests of the tropics, with a rich food supply and high metabolic rate all the year round. Our most characteristic European families are the Helicidae, Limacidae, Arionidae and Clausiliidae; and every warmer region has its recognizable stamp: Africa with its giant Achatinas, South America with large and highly coloured snails more numerous than anywhere else, including herbivorous *Bulimulus* and *Bulimus*, and carnivorous *Glandina* and *Streptaxis*. In Malaya, China, the E. Indies and N. Australia the tropical Helicidae have their headquarters. A peculiar region is New Zealand, with its large fauna of primitive endodontids, large species of carnivorous *Paryphanta* and lack of helicids. Hawaii is the centre of the Achatinellacea, and in the Pacific are also centred the bulimulids, *Partula* in the Polynesian area and *Placostylus* on the islands of Melanesia.

The best modern account of many aspects of the pulmonates is the two-volume symposium, *Pulmonates* (1975), edited by Vera Fretter and J. F. Peake.

Evolution in the Opisthobranchia

Thiele has recognized some sixty-nine families of opisthobranchs, yet this is a smaller group than either the prosobranchs or pulmonates. Opisthobranchs are the most typically marine of gastropods, and the sacoglossans of tide-pools, such as *Actaeonia*, or of salt marshes, as *Limapontia*, are almost the only high-tidal representatives. In evolutionary enterprise, however, the opisthobranchs rank first among the Gastropoda: they are a living museum of adaptive morphology, nearly every family having some distinctive pattern to show. Like the pulmonates the opisthobranchs must have early arisen from a prosobranch stock and the evolution of shelled opisthobranchs was well under way by the end of the Carboniferous. A primitive shelled form like *Actaeon*[138] has much in common with both prosobranchs and early pulmonates, but thereafter the three classes become widely different.

We have already seen the way the opisthobranchs escaped from torsion and dispensed with the spiral shell; and how with the disappearance of torsion the mantle cavity was reduced and eventually lost, along with the ctenidial gill. With the slug-like body, the way was set for a rich evolution of new forms, both bottom-dwelling and swimming. The loss of the shell and gill is the basis of the old division into 'tectibranchs' and 'nudibranchs'. This unfortunately cuts across several natural series, for the opisthobranchs cleave deeply into a number of radial lines, independent almost from the beginning. In the classification adopted here we have recognized eight ordinal divisions, with their separate clusters of lineages running through a series of broadly similar changes. Though not every one of these 'scenarios' is complete, the Opisthobranchia are a good example of what has been styled 'programme evolution'. At the lower reaches of several orders, we find opisthobranchs with a spiral shell, sometimes an operculum, and a mantle cavity with a ctenidium and detorsion only just beginning (Fig. 54).

The Cephalaspidea show some evidence of being themselves a heterogeneous group, with the fully shelled Actaeonidae remote from the rest; *Actaeon* shows, however, some highly primitive features, with an operculum, full torsion including a figure-of-eight nerve loop, and generalized mantle cavity, gut and reproductive system. This can be paralleled among the Sacoglossa by shelled 'bulloid' forms of a similar kind, though without an operculum, such as *Cylindrobulla*, *Arthessa* and *Oxynoe*. The Anaspidea or aplysioids begin with the fully shelled bulloid-like *Akera*. The first of the Thecosomata – the Limacinidae – are spirally shelled, with an oper-culum, though all have lost the gill*, and in the Notaspidea or Pleurobranchoids the earliest members have a large external limpet-like shell. The various groups show different degrees of progress: the Cephalaspidea, the most primitive, seldom lose the shell, while in the 'highest' – the Nudibranchia – it is never present, and the naked body may become very specialized.

Some early steps in opisthobranch evolution are illustrated in the Cephalaspidea (bulloids) and the Anaspidea (aplysioids), the first group chiefly burrowers, the second living mainly on algae at the surface. The bulloids culminate in types such as *Philine* with a flattened wedge-like body, internal shell and strong calcified gizzard adapted for crushing the shelled prey.[172] The later aplysioids are larger and plump-bodied, with a vestigial internal shell and a gizzard employed for triturating and straining algal food. A remarkable side theme in both groups is swimming, best developed by *Akera* among aplysioids, which is very exactly matched by *Gasteropteron* among the bulloids.

The swimming habit is well exploited by the pteropods of the order Thecosomata, generally held to be derived from early cephalaspids. These are of modest size, entirely pelagic and ciliary feeding, being an incredibly numerous component of the plankton. The least modi-fied family, Limacinidae, well reveal how thecosomes may have been derived from neotenic bulloid larvae,[219] they have still a sinistral spiral shell and an operculum, and, apart from its long parapodial wings, the foot retains a wide sole. The Cavoliniidae have acquired bilateral symmetry, with the shell produced into a narrow cone, or a flattened case, variously equipped with spines (p. 47). The Cymbuli-idae show a different trend, with the development of a boat-shaped pseudoconcha; with them should be mentioned the curious genus *Peracle*, which has all the appearance of a *Limacina*, with its coiled shell and operculum, but is in fact probably ancestral to the Cymbuli-idae (Fig. 16).[315]

The other order of pteropods – the Gymnosomata – are some of the most advanced and isolated opisthobranchs. They are all active swimmers, lacking an adult shell, mantle cavity and ctenidium. The larval shell is fragile and thimble-shaped. The body is fusiform, sometimes with secondary 'adaptive gills', and like the heteropods

Peracle is said to have a true ctenidium.

and gymnosomes are rapacious carnivores. They feed mainly on thecosomes, with which they usually swarm, and specialize in large batteries of prehensile buccal tentacles, hooks and suckers. Their buccal armature gives them some of the advantages of both cephalopods and chaetognath worms.[224]

The Acochlidiacea are a small and little-known order of uncertain affinities, all minute and dwelling in the interstitial water space of coarse and finer sand. *Hedylopsis suecica* was, for example, described by Odhner from shell sand, along with *Amphioxus, Caecum* and the minute annelid *Polygordius.* It is about 2 mm long, with the anterior end narrow and extensible for thrusting into narrow spaces. The thin visceral sac projects well behind the foot, being covered only with spicules. The animal – like all the group – is a deposit feeder, shy in habits, hiding in small empty shells, and rolling itself protectively into a ball. Other genera such as *Acochlidium, Microhedyle* and *Strubellia* are receiving many additions of species with the growing study of the interstitial fauna.

The order Sacoglossa we have already characterized (p. 113) by their peculiar suctorial feeding, extracting the cell fluids of green algae by the piercing radula and force-pump pharynx.[120] Their diet commits them to a small size, never longer than about three-quarters of an inch. In the buccal organs the Sacoglossa stand apart from all the rest of the class, but in other characters they seem to follow the opisthobranch 'programme' fairly completely. There are seven families, of which the first two, the Arthessidae and the Oxynoidae, are primitively shelled and ctenidiate, broadly comparable to *Actaeon* or to prosobranchs.

The extraordinary offshoot of the bivalved sacoglossans of the family Juliidae was brought to light only by Kawaguti and Baba in 1959.[179] Living on *Caulerpa*, these small slugs possess a shell that had been previously classed as lamellibranch, distinguished only by a short, sinistrally coiled protoconch on the left valve. Fig. 53 shows the essential sacoglossan morphology of *Berthelinia limax*, and the way the shell develops in the veliger larva, with a second (right) calcification centre. There is a single adductor muscle.

Kay[180] has fully reviewed the radiation of the Sacoglossa. The Juliidae (genera *Julia, Edentellina,* and *Berthelinia* – with subfossils) appear to be a derivative of a primitive stock of shelled Sacoglossa. *Cylindrobulla* and *Volvatella* are primitive burrowing forms, with an adductor muscle running to the flexible outer lip of the shell, corresponding to the separated part in juliids. Other shelled

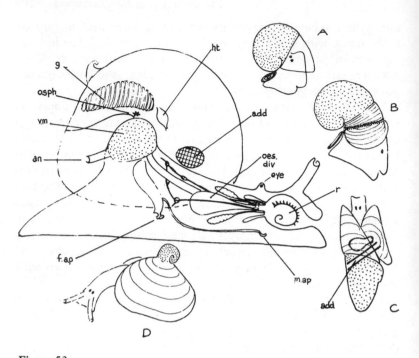

Figure 53
The bivalved sacoglossan *Berthelinia limax*, after removal of the right valve
(*Inset A to D*) Stages in development of the shell
a, anus; add, adductor muscle; e, eye; f.ap, female genital aperture; ht, heart; m.ap, male aperture; oes. div, oesophageal diverticulum; osph, osphradium; r, radula; v.m, visceral mass

forms, *Lobiger* and *Oxynoe*, form an early swimming radiation while the bivalved Juliidae are a creeping stock. The 'nudibranch' sacoglossans are derived from the 'tectibranchs'. In the Elysiidae and Calliphyllidae the body is flattened, with leaf-like lateral expansions. In the Hermaeidae (in tidal rock pools) the digestive gland has ramified into dorsal appendages as in aeoliids. The high-tidal Limapontiidae (*Limapontia*, *Actaeonia* and *Cenia*) have a smooth, narrow, slug-like body.

The Notaspidea are rather generalized opisthobranchs, foreshadowing the true nudibranchs, to which they have almost certainly given rise. In *Pleurobranchus* the body is flat and slug-like, the mantle covering the dorsal surface and projecting as a skirt. From beneath it in front extend grooved rhinophores; on the right side it shelters a true ctenidium and an osphradium, though there is no mantle cavity. In *Tylodina*, the most primitive-looking of all Notaspidea,[10] there is a broad limpet shell, covering the animal entirely. In *Umbraculum* the external shell is circular and sits aloft like a chinaman's hat. The radula is very broad, with numerous, primitive, simple teeth. *Umbraculum* has the record number of 150000 teeth.

The true Nudibranchia are grouped into four superfamilies. They are the dominant section of the opisthobranchs, and the most colourful and extravagant of all gastropods. Detorsion is complete, the shell, gill and mantle cavity are wholly lost, and rhinophores as in other opisthobranchs replace the head tentacles and osphradium, standing up club-like on the front of the dorsum. As we have already seen, each superfamily carried its special stamp. Some trends are pushed ahead in different groups in a roughly parallel way, such as the reduction of the mantle edge and its replacement by various dorsal outgrowths; the adaptation of the naked surface for protective and respiratory functions; and the breaking up of the digestive gland and its deployment in the dorsal body wall. The reproductive organs become very complex, especially in the Doridacea; and the digestive organs become adapted for a grazing carnivorous life, and finally for suctorial habits.[124, 206]

In some ways the Dendronotacea, especially, for example, the Tritoniidae, are the most primitive nudibranchs; but this superfamily has become variously specialized. They are technically distinguished by the rhinophore having a basal sleeve, formed from part of the mantle. A very common feature is the lavish outgrowths of the dorsal surface, such as branching gill tufts in *Tritonia*, or spoon-shaped processes in the Lomanotidae. In the Dendronotidae these processes are richly branched, and in the Dotonidae they resemble cerata, provided with branch tubercles arranged in rings. Among the swimming Dendronotacea are both the large-cowled, heavy-bodied Tethyidae reaching 30 cm (12 ins) long (p. 45) and the delicate pelagic Phyllirhoidae.

The hallmark of the Aeolidiacea is the habit of feeding on coelenterates and storing their nematocysts in cnidosacs in the cerata. With this means of protection they may develop very bright warning coloration. The cerata may be tapered or club-shaped, and develop various patterns of arrangement. *Coryphella* is a rather primitive British form with short cerata. *Facelina* and *Aeolidia* are other well-known genera. *Glaucus* and *Fiona*, which feed on siphonophores, are pelagic Aeolidiacea; and one family, the small worm-like Pseudo-

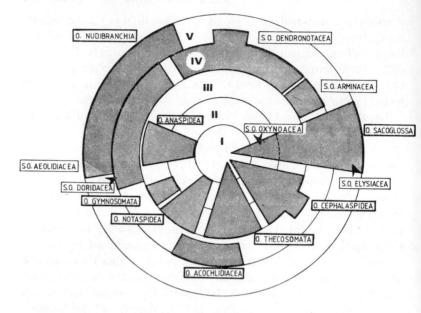

Figure 54 Radial classification of the Opisthobranchia, suggesting the separate course of the evolutionary programme in the class, with the different orders passing through five grades of organization:
I Primitively with full shell
II Body becoming slug-like and shell reduced
III Slug facies completed and shell only a vestige (Gymnosomata naked and swimming)
IV Shell lost and nudibranch bilateral symmetry
V Digestive gland re-deployed in external cerata

vermidae, is held to represent simplified aeoliids probably living on sand-dwelling interstitial hydroids.

In the superfamily Doridacea the body is typically flat and the foot wide. The anus lies posteriorly at the middle of the upper surface, and around it is a circlet of pinnate secondary gills. The mantle is often provided with spicules, but the digestive gland does not ramify in it and cerata are never found. Early Doridacea, such as *Gnathodoris* and *Bathydoris*, are probably derived from tritoniids, and like them have jaws and a broad radula, being grazing carnivores or deposit scrapers. The Dorididae (*Archidoris*, *Jorunna*, etc.) have a similar broad radula, and graze on sponges. They have also a wide mantle brim. Higher families show two trends: the retreat of the mantle brim, with its replacement sometimes by tentacular processes, and the narrowing of the radula (sometimes its total loss) with the development of suctorial feeding (p. 111). In the Notodorididae the mantle is a good deal reduced, and in the Polyceridae (*Polycera* and *Triopa*) long marginal 'horns' develop. The Onchidorididae (*Onchidoris* and *Acanthodoris*) and the Goniodidoridae (*Goniodoris* and *Ancula*) are suctorial feeders with buccal pumps, and in the latter family the mantle brim is narrow or lacking.

The best-known British representative of the Arminacea is the yellow and black striped *Pleurophyllidia*, with secondary gill-lamellae crowded beneath the mantle skirt. The tropical *Phyllidia* feeds in the same way on sponges. The North American *Armina* is adapted for shallow burrowing. It feeds above the sand on sea-pens (*Ptilosarcus*), using the strong jaws and radula as does *Tritonia* on soft corals. The digestive gland has penetrated into small folds like cerata. Revision of this mixed group may be needed, as our knowledge improves.

Parasitic Gastropoda

Most of those gastropods that have become parasitic have been generally placed in the prosobranch order Mesogastropoda. They show two very different types of habit: some are ectoparasites with a normally shaped body and a long spiral shell, others have adopted a more permanent attachment to the host, which may culminate in complete endoparasitism, with the normal molluscan characters quite discarded. The most advanced members of this group are even more simplified than tapeworms. It is only by their veliger larva, with a spirally coiled horny shell, that we can recognize them as molluscs at all (Fig. 55).

To the first group belong two main families, the Eulimidae and the Pyramidellidae.[137] The eulimids are ectoparasites of echinoderms. The shell is usually smooth and highly glazed, and sometimes – as in the British *Balcis devians* (Fig. 55), a parasite on *Antedon bifida* – the tip of the spire slants a little to one side. The Pyramidellidae, like the Eulimidae, are generally tiny, hardly ever more than 18 mm in length, with slender shells, often ribbed or strongly sculptured. They associate chiefly with bivalve molluscs, sedentary worms and coelenterates, crawling over or hanging by the proboscis from the soft body of the host. They insert the slender proboscis into the body wall, sucking blood or tissue fluids – in *Turbonilla interrupta*, for example, from the mantle edge of bivalves, in *Turbonilla elegantissima*

G

from the blood gills of cirratulid worms. Both eulimids and pyramidellids are fairly host-specific. Though usually overlooked by collectors they can often be taken plentifully by careful searching in the neighbourhood of the known host. Fretter and Graham have added much to our knowledge of British species; the following is a list of some pyramidellid parasites and their hosts.[137, 12]

Odostomia rissoides	*Mytilus edulis*
O. scalaris	*M. edulis*
O. lukisii, O. unidentata, O. plicata	*Pomatoceros triqueter*
O. eulimoides	*Pectinidae*
O. trifida	*Mya arenaria*
Chrysallida spiralis	*Sabellaria spinulosa*
C. obtusa	*Ostrea edulis*
Turbonilla interrupta	*Ostrea, Pecten, Venus*
T. elegantissima	*Cirratulus, Audouinia, Amphitrite*
T. jeffreysi	*Halecium* and other hydroids
Angustispira spengeli (Indo-Pacific)	*Meleagrina*

Both these families were until recently placed in the mesogastropod superfamily Aglossa, but, as we have seen (p. 105), there are fundamental differences in their feeding organs. The Pyramidellidae have chitinous jaws exquisitely adapted as piercing stylets, while the Eulimidae have an unarmed proboscis, evidently softening the host tissues by secreting enzymes. The Eulimidae – and perhaps indirectly the Pyramidellidae – are descended from the less modified family Aclididae, with long-spired shells but only moderately long proboscis and normal jaws and radula.

Fretter and Graham believe that the pyramidellids are not prosobranchs at all, but opisthobranchs specialized at an early level. Though the external structure gives little hint of it, there is good evidence for this view. The Pyramidellidae appear geologically at about the same time as the Acteonidae (Carboniferous) and much earlier than most mesogastropods. They show many quite circumstantial resemblances to opisthobranchs, of a kind that are unlikely to be adaptive, and can thus be more reliably used as taxonomic features. As in early opisthobranchs and pulmonates as well, the larval shell is at first sinistrally coiled, later reversed to dextral. The shell aperture, the tentacles and the structure of the hermaphrodite genital system and of the gut point to opisthobranch connections. The same holds good – as was found recently – for cytological details of the sperm. If, in spite of first appearances, we are to rank the pyramidellids with opisthobranchs, we shall need a separate order, quite distinctive from bullomorphs and showing some likenesses to prosobranchs. At this point in the classification, the three sub-classes Prosobranchia, Opisthobranchia and Pulmonata undoubtedly draw very close together.[220]

In the ectoparasite *Thyca*, which lives permanently attached in the ambulacral grooves of starfish, the shell is depressed and cap-shaped, the foot reduced, and the proboscis, with neither radula nor jaws, is permanently plunged through the integument of the host. *Thyca* is generally classed with the Capulidae. From here, or from Eulimidae, may have arisen the next family – Stiliferidae – which are still more intimate parasites of echinoderms. The first genus, *Mucronalia*, has a

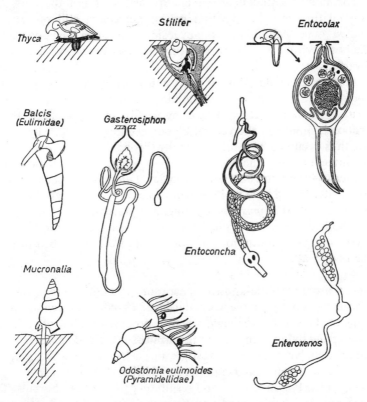

Figure 55 Ectoparasitic and endoparasitic Gastropoda
Balcis (after Fretter); Odostomia (after Fretter and Graham); (others after various authors, from Baer)

long spiral shell like the eulimids, a vestigial foot with a small operculum, and a proboscis striking deeply into the host – urchin, brittle star, starfish or holothurian. At the base of the proboscis is a fleshy frill, which in *Stilifer* is destined to grow up round the spiral body of the animal to form an enclosing sac, the *pseudopallium*. The parasite now sinks deeper into the host (a starfish or urchin) and the flask-shaped pseudopallium opens by an external pore through which respiratory water is pumped. In the pseudopallial cavity lies the spiral body with its horny shell. *Gasterosiphon* is completely internal. Its swollen pseudopallium still conceals a small visceral mass and communicates by a narrow siphon with the world outside. Most of its length – about 12 cm – consists essentially of a tubular proboscis lying freely in the body cavity of a holothurian.

The family Entoconchidae, endoparasites of holothurians, has evolved further still. The first genus, *Entocolax*, still possesses a dilated pseudopallium opening by a siphon through the body wall of the host. Within it lodge several tiny cellular spheres, which are the dwarf males, consisting of little more than a testis. The visceral mass of the female is reduced to an ovary and short oviduct, the rest of the body being a long proboscis containing the simplified gut. *Entoconcha* is a worm-like creature measuring 7·5–10 cm long, fixed by its proboscis mouth to the ventral blood vessel of its host. Most of its length is occupied by a tubular cavity (probably a pseudopallium in origin) containing the eggs, and in a slight terminal swelling a cluster of fifteen to twenty dwarf males. *Enteroxenos* reaches a final simplicity. It is a pale white worm up to 15 cm in length, attached to the gut of *Stichopus* when young, afterwards free in the coelom. The body cavity is a spacious uterus filled with developing eggs, and there is neither mouth nor gut, food being freely absorbed through the body wall.

There are two little-known genera of parasitic mesogastropods placed by some near the Lamellariidae: *Ctenosculum* living on the arms of the starfish *Brisinga*, where it produces a kind of gall, and *Asterophila* living immersed in the arm tissues of *Pedicellaster*. Each has a globular body, with vestiges of gastropod organization, enclosed in a pseudopallial capsule. In addition there is *Paedophoropus* living in the Polian vesicles and respiratory tree of holothurians. The proboscis is very enlarged and the female has a dilated brood sac formed by the foot.

10 The Bivalvia and their classification

Though the bivalves are wonderfully diverse in form and habit, their basic pattern is unmistakable. No bivalve has a head, buccal mass or radula. In nearly all of them the mantle encloses the whole body, and is itself covered by a two-piece shell.* The role of food-catching has passed from the head to the gill, and the main sense organs have moved to the edge of the mantle. In higher forms the mantle cavity may be closed to the world outside, except for the siphons and the foot-gape, and the animal burrows deeply into the substrate. Great emphasis is placed on mucus and cilia in the life of bivalves, and the pallial organs and the stomach are especially elaborate. The chief muscular organs – besides the foot – are the all-important shell adductors.

Of the first and oldest group of bivalves, the Protobranchia, only a few living genera remain, all – as we have seen – rather specialized. The Nuculidae and Malletiidae, for example, feed on surface deposits by means of peculiar palp proboscides (p. 89). *Malletia* and *Yoldia* have highly enlarged pumping ctenidia. The Solenomyidae, with their long, tubular, mainly periostracal shell, have developed a power of darting and swimming with the piston-like foot. But in spite of these peculiarities, which the most primitive ancestors of lamellibranchs could not have possessed, palaeontology and comparative morphology agree in pointing to the protobranch bivalves as the forerunners of all the rest. The Nuculidae already existed in the Silurian and

---

*Dr N. J. Morris, of the British Museum, has directed some valuable light on the possible origin of bivalves from single-piece ancestors near hypothetic Monoplacophora. The ribeirioid bivalves from Cambrian to Ordovician (*Ribeiria* and *Tecnophorus*) – homeomorphic in shape with nuculanid prosobranchs – have a laterally compressed, two-sided shell as in other bivalves, but the shell structure retains a complete continuity overarching the median line. In *Conocardium*, first found in mid-Ordovician, where the shell gapes anteriorly for a presumed burrowing foot and has a long rostral tube posteriorly, the dorsal continuity of the shell is still maintained by the nacreous layer, but erosion takes place externally to develop a potential hinge along this line.

related shells are found in the Cambrian. In living forms the ctenidia, arrangement of the mantle cavity, and structure of the gut and reproductive organs, have an indelible primitive stamp. In all their characters – whether primitive or specialized – protobranchs stand clearly apart from other bivalves; on many grounds they deserve to be separated into their own sub-class.

Above the Protobranchia we run into difficulties in classification. With such a constant basic structure we shall expect to find much parallel evolution in lamellibranchs, and there are three different types of character of which we must take account. First, there are 'adaptive' characters, which are immediate modifications for a particular mode of life, and thus dangerous to use in classification. Secondly, there are 'progressive' characters, those which usually show a definite trend of advance, often running parallel through several unrelated groups; and thirdly, 'conservative' characters – those that persist unchanged or stable over long periods. It is the last that will be most useful in showing real affinities.[229]

There have been traditionally two main bases of classification of lamellibranchs, employing either gills or the hinge characters, neither by themselves fully satisfactory. Pelseneer's system (1889) attached particular importance to the condition of the gills, whether 'protobranch', 'filibranch', 'psuedolamellibranch', 'eulamellibranch' or 'septibranch'. To lay stress on these characters Pelseneer made them the basis of his five orders. In 1906 he discarded the order Pseudolamellibranchia, distributing some families to the Filibranchia, others to the Eulamellibranchia. In 1911 he revived this order. He failed to realize that the condition of the gill is a progressive character, passing through successive stages of evolution in several parallel lines. It leads to a 'horizontal' classification rather than a natural or 'vertical' grouping.

European workers such as Thiele (1935)[11] and American and British conchologists, such as Winckworth in his British List, have in part replaced Pelseneer's system by drawing on various classifications that use the structure of the hinge, such as that of Cossmann (1914) and Dall (1895 and 1916). This has the advantage of keeping the palaeontologist in harmony with the zoologist. Its great weakness is in the neglect of soft parts: for example, the obviously distinct sub-class of protobranchs is placed in one group with some or all of the filibranchs.

In later times Atkins was to suggest regrouping based on the ciliation of the gills, and Purchon has put forward a classification with new ordinal names on the basis of internal stomach structure. Most single-based systems break under strain at some points: what is needed is a liberal classification blending evidence from as many reliable sources as possible. One such arrangement was attempted by the French conchologist Douvillé,[108] who sketched a provisional classification without proposing any formal names. He was the first to emphasize that in such a variously evolving class, the plasticity of each line may lead frequently to convergent cross-resemblances. His arrangement was almost overlooked until brought to notice again in 1932 by Davies in a stimulating essay on bivalve classification.

Douvillé's classification of lamellibranchs struck an ecological note, beginning with three parallel streams, normal, sessile and deep-burrowing. Each line formed in turn, like a many-stranded rope, the strands in general advancing together but fraying out repeatedly into adaptive radiations. Douvillé was aware of pitfalls of convergent cross-resemblance, but underestimated them: with the insistent adaptive variation and mosaic evolution among the lamellibranchs the broad lineages become deeply and completely divided.

How may we organize this rich material today? First, among what used to be Pelseneer's Filibranchia and Pseudolamellibranchia, we can recognize a recurring set of characters. Attachment by the byssus or secondarily by shell cementation is the leading habit: only a few forms are free-moving. The shells are rather thin and often highly pigmented. The nacreous as well as the prismatic layer is highly developed.

The remaining bivalves form a large assemblage with many parallel trends but all agreeing in increasing commitment to burrow, with consequent fusion of mantle margins affecting also siphon and ligament. More generalized members have solid white porcellanous shells and usually burrow shallowly, as with the plump *Cardium*, *Venus* or *Astarte*. Further adaptations to deep burrowing involve continuation of the same trends, extension of fused siphons, lightening and streamlining of the shell and sometimes reduction of the foot. This has produced modern bivalves with long fused siphons, a closed mantle cavity, and a gaping shell with poorly developed hinge. Such forms include the immobile *Panopea* and *Mya* and the piddocks (Pholadidae) which burrow into firm substrates.

1 Surface-attached lamellibranchs
Orders Taxodonta and Anisomyaria

These, after the prosobranchs, include the oldest bivalves. They became fashionable in the middle and later Palaeozoic and early Mesozoic, when most of the living families were established. The majority are filibranchs but *Ostrea* and *Lima* develop pseudolamellibranch gills (see p. 90). They vary greatly in appearance but all agree in the widely open mantle and lack of siphons, the importance at some stage at least, and generally throughout life, of the byssus (though the oysters lose it after settlement), and the tendency to asymmetry between the anterior and posterior parts of the body. In an old but still useful study, Jackson (1890) united the 'Avidulidae and their allies' with the Arcidae and the constituent families have been loosely grouped ever since. L. R. Cox in a recent palaeontological classification recognizes four orders (Eutaxodonta, i.e. our Taxodonta; Isotilibranchida for the mytilids; Colloconchida for the oysters; and Pteroconchida for the rest). All four are usefully embraced in the sub-class Pteriomorphia.[91]

In all but the earliest of this series the primitive symmetry is greatly modified (Fig. 56). The most ancient superfamily is the Arcacea noah's ark shells and their relatives, which may be placed alone in the order Arcoida. These have *taxodont* dentition, a long row of uniform teeth as in *Nucula*. The two adductor muscles are equal and the posterior halves of the body approximately equally developed, as in the British *Glycymeris*, which has the rounded shape of a venusshell. *Glycymeris* and *Limopsis* burrow shallowly but *Arca* rests at the surface with the dorsal side uppermost, attached by a byssus springing from the whole edge of the foot.

In the remaining families (order Mytiloida, and more so in the Pterioida), the sessile habit has caused profound changes.[329] With the mussels (Mytilidae) the byssus and foot have moved to the anterior end, restricting the anterior adductor muscle to a small size (the *heteromyarian* condition). In *Modiolus* and *Mytilus* the animal is still attached upright to the substrate, and – especially in *Mytilus* – the anterior end is small and pointed, the posterior end broad and rounded, with a large posterior adductor. The Mytilidae show considerable adaptive radiation. The genus *Modiolaria* contains small nut-shaped bivalves, embedded in a nest of their own byssus threads or in ascidian tests. Two genera have become narrow and elongate, burrowing in rocks. *Botula* is attached by byssus threads within its burrow in soft non-calcareous rocks, the ridged shell valves abrading the rock by opening thrusts of the ligament. *Lithophaga*, the date mussel, bores by rotating the shell in calcareous rocks; the fused inner lobes of the mantle are glandular and appear to secrete an acid mucus.[331]

In more advanced Pterioida the shell usually lies upon its right side, becoming flattened, with the two valves no longer quite alike. The pearl oysters, (of the superfamily Pteriacea), have a short byssus emerging by a deep notch in the right (lower) valve. The anterior adductor muscle has been lost, the foot is very small and functionless, and the whole anterior half of the animal is now of minor dimensions. The single posterior adductor muscle lies at the centre of the valves, in the *monomyarian* condition. The mantle cavity lies widely open and water enters around two-thirds or more of the rounded margins of the shell. The pearl oyster itself (*Pinctada*) has almost circular valves, but with a long straight hinge. Later genera, such as the wing shells (*Pteria* or *Avicula*) show a strong tendency to elongate along the hinge-line. In the Vulsellidae – by the same emphasis of the hinge – the exhalant point is carried posteriorly on a long salient, until in the hammer-oysters (*Malleus*) the shell finally becomes T-shaped or hammer-like.

The fan shells, Pinnidae, also belong to the Pterioida, but show a quite different trend of evolution. They have long and wedge-shaped equal valves, and are unique in being embedded upright in the sand and secured there by a byssus. Each byssus thread is attached to a sand particle, and the whole structure gives great stability in a soft substrate. The fan shell is immobile and the foot and anterior end are greatly reduced. The posterior or uppermost part of the shell is broad and triangular, composed of horny conchiolin, only thinly calcified. There is a wide mantle gape at the broad end, with thickened lips, and – as in other Pteriacea – there are efficient ciliary and mucous tracts for cleansing the mantle cavity of sediment. The greater part of the mantle in *Pinna* is free of attachment to the shell, and its edges can be deeply withdrawn and protected from injury by special pallial retractor muscles.[330]

Like the pearl oysters, the Pectinacea or scallops (Fig. 57) lie upon the right side.[329] Both valves may be concave and similar or the left one may be flat. Here, however, the valves have again become symmetrical about the hinge and beak. In the viscera distortion resulting from anterior fixation by the byssus remains, but the shell and mantle have been emancipated from the rest of the body, and – unlike the

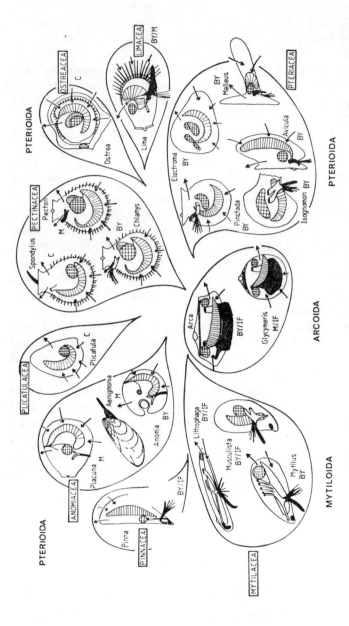

Figure 56 Series of anisomyarian lamellibranchs, showing the development of the various monomyarian lines with the loss of the byssus and the development of free mobility, or cementation
BY byssus-attached; C, cemented; IF, infaunal; M, mobile, including swimming

Pteriacea – have a new superficial symmetry. An early representative such as *Chlamys varians* still shows a slight asymmetry, with a byssus as in *Pteria* passing through a notch in the right or lower valve, and the squared 'lugs' at either side of the hinge unequally developed. The exhalant current is expelled at these points, immediately to the sides of the narrow hinge. Incoming water can thus enter right round the rest of the circumference and flow straight through to the hinge side. Attachment by the byssus may soon be lost, as in the massive rock scallop, *Hinnites*, and the thorny oysters of the related family, Spondylidae, in favour of direct cementation. Such forms lie in a deep cup-shaped right valve. Though the mantle is widely open to sediment, the silting up of this valve is prevented by the great development of the 'quick' part of the adductor muscle. The shell can be rapidly opened and closed and the sediment falling on the mantle edge so removed. In the Pectinacea the mantle edge carries a profusion of sense organs, with tactile tentacles sensitive to sediment and complex eyes (p. 165).

Pecten itself was never cemented, but on gaining freedom from the byssus developed the swimming habit. For this – as in *Chlamys opercularis* and still more so with the thin circular shell of the related genus *Amussium* – the body is admirably adapted. The 'quick' muscle forms a mechanism for clapping the valves, and the pallial margin with its light sense and velar curtain controlling the exit of water are important preadaptations for swimming (see Fig. 58).

The Limidae or file shells are classified near the Pectinacea. Some species such as *Lima excavata* are attached by the byssus from which they may spin a nest of dense threads. In some species the gills and mantle are scarlet and the long pallial tentacles white. In *Lima hians* the tentacles are long and complex, brilliantly orange. They produce a sticky defensive secretion and may, by thrusting into gravel, assist the animal in embedding itself. Alternatively, *Lima hians* may swim as *Pecten*, but always with the ventral margin foremost, taking a bite out of the water and expelling water postero-dorsally. The very flexible tentacles take part in swimming by an apparent rowing action, with effector and recovery strokes in analogy with cilia.

In the crawling movement of *Lima* the foot is uniquely reversed, protruding between the rounded margins so that the anterior (hinge) end moves hindermost.

The most remarkably altered of all byssus-attached bivalves are the saddle oysters, Anomiacea. In *Anomia* and *Monia* the byssal notch seen in *Pteria* and *Chlamys* is now very deeply embayed in the right valve. The byssus threads coalesce into a calcified cable, which emerges near the centre of this thin lower valve. The upper valve is convex and moulded to the substrate, producing a strong bilateral asymmetry. The byssal retractor muscles are attached to the upper valve only, and serve, like the shell muscles of a limpet, to pull the animal down against the substrate. There is thus virtually only one functional valve, fitting close to the ground like a limpet shell. Since this is adducted by the byssal muscles, the true adductor is tiny and without function. The tropical saddle oyster, *Placuna*, has lost its byssus and lies freely on the ground. Uniquely it has turned over to lie on the concave left valve, whose upturned margins raise the edges of the mantle clear of sediment.

The evolution of the Anomiidae has not been exhausted by the fixed posture with calcified byssus cable. Owen and Yonge have recently described the habits of *Aenigmonia*, a small elongate limpet-like mollusc, creeping by means of its long extensile foot upon mangrove leaves in Malaysian estuaries. When closely examined, it is found to be a true anomiid that has abandoned its byssus attachment. The lower valve is almost functionless and the foot has recovered its free locomotor powers.[332]

The oysters, Ostreacea, are invariably fixed and sedentary. They lose the byssus at the spat stage, and the young become cemented by the right valve immediately upon settling. The foot is lost entirely, almost uniquely in bivalves, and the crescentic gill extends round a wide part of the mantle circumference. The 'quick muscle' is highly adapted for expelling sediment; and the low or sub-tidal oysters of the genus *Ostrea* lie frequently upon silty shell or gravel ground. Typified by European *O. edulis*, these oysters are incubatory, retaining the eggs through to the larval stage, and reach higher latitudes; the oviparous oysters, *Crassostrea*, are a warm-temperate group and are frequently mid-tidal and zone-forming upon rocks or piles.[24]

2 Shallow-burrowing lamellibranchs
Orders Heterodonta and Schizodonta

The bivalves of the 'eulamellibranch' series have avoided the extreme specializations due to attached life. Yet they are not to be thought of as primitive, and have on the whole flourished later than the attached forms, in later Mesozoic, Tertiary and Recent times. A majority of species lie freely at or near the surface, or burrow actively in sand or mud. An insistent trend is the tendency to closure of the mantle and development of siphons. Though in such a large group, any general statement will oversimplify, their earliest habit is probably the shallow burrowing we have met with already in the cockles (Cardiacea) and the venus-shells (Veneracea). Two such early groups, with the mantle widely open and the siphons short or wholly lacking, are the Astartacea, well represented in Britain, and the Carditacea. Here the primitive genus is the cockle-like *Venericardia*, while *Cardita* is attached by a byssus, and has become 'mytilized' or heteromyarian, with the posterior end much expanded.

In the plump, rounded cockle-like forms often with solid shells, the foot is sometimes a powerful muscular tongue, and the animal leaps (p. 57). This may be so in some of the Cardiacea or cockles, a large group in tropical and temperate seas with a great variety of radial or spiny sculpture. *Corculum cardissa* is an extraordinarily modified Indo-Pacific cockle, flattened anteroposteriorly, and broadly heart-shaped viewed from behind. In Britain there are the Cyprinacea (with one species, *Cyprina islandica*) and the Isocardiacea; *Isocardia cor* has a large orbicular shell with the umbones spirally coiled. The Veneracea or venus shells are a highly successful group with a considerable form range. In shallow-burrowing species, as of *Venus*, *Chione* and *Antigona*, the siphons are short and the shell cockle-like, though with the predominant sculpture not radial but concentric, and sometimes lamellate. The rectangular *Paphia* and *Tapes* burrow rather more deeply and the siphons are longer, while in deep-burrowing, rounded *Dosinia* they are very long and lightly fused.[54]

The superfamily Mactracea, the trough shells or surf clams, are the characteristic sand-burrowers of exposed shores. Their shells are rather large, smooth and light and oval to elongate. The two siphons are long and fused, and the mantle lobes well united. They may also, like the cockles, use the flexed foot in leaping.

The Lucinacea are a large group of round-shelled bivalves, mainly burrowers in tropical sandy mud and sea-grass flats. They are primitive in having no extensive fusion of the mantle edges, and two of their families – the fragile-shelled Ungulinidae and Thyasiridae – have no siphons. The more solidly built Lucinidae have an exhalant siphon which is unique in having no siphonal retractor muscles and no pallial sinus. Retraction involves simply shortening and inversion. The Lucinidae have, in addition, unique respiratory folds of the mantle-lining epithelium. In all Lucinacea, the foot, in addition to burrowing and locomotion, constructs the anterior inhalant shaft through the sand. In the Lucinidae it is long and vermiform, up to six times the shell length, and secretes mucus to bind the sand.[44]

The most adept of burrowing bivalves are probably, however, the Tellinacea, where – as we have seen – the shell becomes thin, smooth and narrowly compressed from side to side. The foot expands to a wide, thin-edged blade, and the separate siphons are very long. While most bivalves are suspension feeders by filtering off particles from the water current, the Tellinacea have become specialized for feeding on surface deposits (p. 93). The sand-dwelling Tellinidae themselves have fragile and rather elongate shells, often shining and delicately coloured. The most beautiful of bivalve shells, thin and wafer-like and rayed in pinks and mauves, or with golden epidermis, are the sunset shells of the Asaphidae. They are fast burrowers and in *Solecurtus* and *Tagelus* run parallel with the razor shells, Solenacea. Also of the Tellinacea are the wedge clams, Donacidae, of mid-tidal beaches, small, solid and triangular, often handsomely and variously coloured. The heavy, triangular freshwater donacid, *Egeria*, is an important item of fishery in West African rivers.

All the above families are members of the order Heterodonta. The hinge teeth are well developed and important in classification, forming well-marked sets of cardinals and laterals. We may consider now some of the more specialized off-shoots which the heterodont lamellibranchs of the normal branch have produced.

First, in the superfamily Erycinacea, are some bivalves of very small size. Their primitive habit is to nestle in crevices, as with *Lasaea rubra*, reaching to high water of spring tide, and *Kellya suborbicularis* at low tide. In some way they recall nuculoids in habits, crawling freely on the surface, taking in their inhalant current by a special anterior siphon and expelling it behind. More advanced members of the Erycinacea show leanings to commensal and even parasitic life. The foot may reacquire a flattened sole and the shell becomes thin and transparent, gaping widely and partly enveloped by the mantle. Such bivalves typically frequent the burrows of marine invertebrates, such as crustaceans and particularly sand-dwelling echinoderms. There are a number of such commensal pairs in the British fauna, the bivalve always very specific to its chosen host. *Devonia perrieri* lives with the holothurian *Leptosynapta inhaerens*, *Mysella bidentata* with the sand-dwelling brittle star *Acrocnida brachiata*. *Montacuta substriata* is found in the neighbourhood of the heart urchin *Spatangus purpureus*, and *M. ferruginosa* with *Echinocardium cordatum*. All these appear to feed in the normal way, collecting food by the gill. They may temporarily attach to the host with the byssus, but the exact economy of these partnerships still needs careful study.[141, 242]

The British *Lepton squamosum*, with its very compressed shell, lies against the wall of the burrow of the crustacean *Upogebia stellata*. A Japanese commensal, *Peregrinamor oshimae*, attaches beneath the body of another *Upogebia* species. Its ventral surface is drawn close to the host by byssus threads, and the shell is peculiarly flattened dorsoventrally, making it heart-shaped in dorsal view. The burrows of the Australian prawn *Axius plectorhynchus* harbour no fewer than six species of commensal bivalves. One of these, *Ephippodonta macdougalii*, keeps its valves permanently agape at 108°. It clings to the substrate by the fused mantle edges, which resemble the foot of a limpet but with a fissure for the true foot to emerge. *Conchenoptyx* and *Coleoconcha*, as described by K. H. Barnard, also hold the shell agape and use a pedal sole.

In *Chlamydoconcha* the mantle wholly surrounds the shell, leaving narrow siphons in front and behind, while *Entovalva* is parasitic – or perhaps inquiline – in the gut of synaptids. Unlike gastropod parasites it is not suctorial, appearing to absorb ready-digested food through the gill and mantle.

The true giants of the lamellibranchs are the clams of the family Tridacnidae, from the tropical Indo-Pacific coral reefs. These are a surface-dwelling offshoot of the Cardiacea or cockles and have become very specialized in their structure and nutrition. *Tridacna gigas* may be 90 cm (3 ft) long and weigh 135 kg (300 lb). *Tridacna fossor* embeds in coral fragments, while the smaller *Tridacna crocea* actively bores. The genus *Hippopus* differs in having no byssal gape. These clams live in the shallow waters of the reefs, and, like some of the Alcyonacea, owe their great success to augmenting their food supply by farming immense numbers of zooxanthellae, or unicellular symbiotic algae.[328] As Yonge has shown, the symmetry of the shell and body is radically altered (Fig. 57). In *Tridacna* a strong byssus emerges through a gape on the lower side of the shell valves which rests close against the ground. The dorsal and ventral aspects are very likely to be confused, since the umbones and hinge have migrated through a full 180° to lie on the ventral side close to the byssal gape. The up-facing open side with the interlocking toothed margins lies in fact – by reference to the viscera – at the dorsal aspect. The anterior adductor muscle has disappeared, and the siphons have migrated upwards from the posterior end to fill the whole dorsal gape. The siphons stay fully expanded in sunlit waters, and their lips are enlarged and fleshy, superficially pigmented with green, brown and yellow to form a screen protecting the symbiotic algae from too intense light. The algae lie in the deeper siphonal tissues, in blood sinus, enveloped by amoebocytes. Light for photosynthesis is focused upon them from groups of hyaline organs acting as lenses, and derived from the well-developed siphonal eyes of the typical *Cardium*. The products of photosynthesis become available after release to the bivalve from the zooxanthellae. The kidneys are abnormally large, for getting rid of the waste products of this digestion, carried to them by amoebocytes. Notwithstanding this

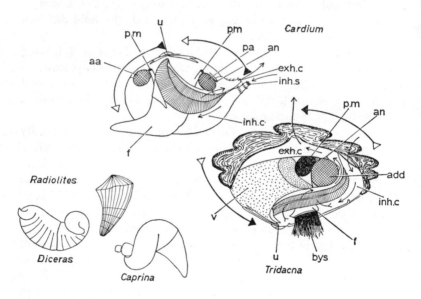

Figure 57   Specialized sedentary eulamellibranchs
(*lower left*) Shells of Mesozoic Chamacea
(*above*) Generalized *Cardium*, to show derivation of organization of *Tridacna* (*right*)
a.a, anterior adductor; an, anus; add, single surviving adductor; bys, byssus; exh.c, exhalant chamber; exh.s, exhalant siphon; f, foot; inh.c, inhalant chamber; inh.s, inhalant siphon; pa, posterior adductor; p.m, pedal muscle; u, umbo; v, visceral mass

special source of food, the Tridacnidae retain typical gills, stomach and crystalline style, and normal ciliary feeding is obviously possible as well.[328]

The superfamily Chamacea[337] have adopted a permanently attached mode of life, cemented upright by the deeply conical right valve, to which the left valve, usually smaller, forms a lid. The Jurassic and Cretaceous representatives are extremely numerous, known chiefly from the large conical Rudistae, such as *Hippurites*. There is only one surviving genus, *Chama*, an oyster-like bivalve, with the foot greatly reduced, and a flat opercular valve. It shows a strong parallel with the gastropod *Hipponyx* (p. 105), which has reached a similar end from very different beginnings. The most ancient Chamacea, the family Diceratidae of the Jurassic, had equal right and left valves, both spirally coiled like ram's horns. In the derivative from *Requiena* the free valve has become operculiform, while *Chama* has been derived in the same way from *Matheronia*. The rudistid family Radiolitidae, was derived from the Caprinidae, with *Caprina* showing the free valve much larger than the fixed and spirally coiled. *Caprinula* shows an intermediate stage in which the fixed valve has much elongated. As in the fossil oysters, *Gryphaea*, orthogenesis seems to have been rife among the Chamacea, and a flourishing Cretaceous stock has left behind it only a single living genus.

Lamellibranchs have spread into lakes and rivers along several separate lines. These have radiated rather little in basic structure, but with geographical isolation have undergone great speciation. The Unionidae, the largest freshwater family, containing the familiar *Unio* and *Anodonta*, are important if only for containing eighty-five genera and some 1000 species, accounting for one-fifth of all living lamellibranchs! We have already referred to their peculiar life histories, with parasitic glochidia larvae. Their three sub-families are the Lampsilinae, with *Lampsilis* and other genera chiefly N. and C. American, the Anodontinae of Europe, America and E. Asia, and the Unioninae, mainly Asiatic and E. Indian though found also in America and Africa. A second family is the Mutelidae, with *Diplodon* and *Hyridella*, the freshwater mussels of the southern hemisphere. The small family Etheriidae has three genera of specialized freshwater 'oysters' in the Amazon basin, tropical Africa and India. *Acostaea* is monomyarian, becoming cemented by the right valve, after which the posterior end alone continues to grow, the anterior end with the umbones remaining as a claw-like appendage.

The above three families form the superfamily Unionacea, belonging to the order Schizodonta. Classification is based on the hinge teeth, with two-limbed Λ-shaped teeth, arranged inside each other in a fan shape. In some Unionidae – such as *Anodonta* – they are lost; but they are well shown in the related Trigoniacea, a filibranch group chiefly found fossil, but with a few living species in the Australian genus *Neotrigonia*.

The remaining freshwater lamellibranchs are chiefly heterodonts. Most successful – after the unionids – are the Sphaeriacea, with the two families Corbiculidae and Sphaeriidae. The world-wide genera *Sphaerium* and *Pisidium* are well represented in Britain, small colourless orbicular bivalves, usually less than a centimetre long, found burrowing shallowly in nearly all types of freshwater habitat. They have short siphons, paired in *Sphaerium*, fused in *Pisidium*, and the mantle lobes almost unfused. The Cardiacea have established a freshwater line in *Adacna* and *Diadacna*, the cockles of the Caspian Sea. In *Dreissensia* we have a mussel from the Caspian Sea (allied by some to *Mytilus*) which has acclimatized itself in English rivers, and still possesses veliger larvae. The W. African *Egeria* is a river-dwelling donacid, *Gnathodon* a brackish water mactrid of the Gulf of Mexico, *Glaucomya* an E. African freshwater razor shell and *Nausitoria* a freshwater Teredo of the Ganges.

3　Deep-burrowing and immobile lamellibranchs
Order Adapedonta

The strongest evolutionary force in this order has been to modify the shell for deep penetration, often with permanent sacrifice of mobility. The mantle is always extensively fused, the hinge is weak with its teeth degenerate or unrepresented. The valves generally gape freely at either end. The immobile habit is frequently obvious from the shell, which tends to be thin and fragile with nothing of the stream-lining or solidity of the Heterodonta.

In the superfamily, Myacea, the deep-burrowing clams *Panopea*, *Mya* and *Platyodon* have a very small foot, together with long fused, leathery-cased siphons and a closed mantle cavity. *Panopea* is generally held to be a primitive deep-burrower, *Mya*, with its asymmetric hinge, to have re-adopted a burrowing habit after a recumbent period upon the surface; *Platyodon* burrows in stiff mud by rocking the shell valves sideways to abrade the wall of the burrow; *P. cancel-*

latus, a Californian myid, continues throughout life to bore efficiently in hard clay. *Mya*, with vestigial foot, becomes finally stationary. The Myacea show a certain adaptive radiation. *Sphenia*, like *Saxicava* (in the nearby Saxicavacea), has short siphons and nestles in shallow crevices. It is also becoming 'mytilized', or heteromyarian, by byssal attachment. One small American species of *Cryptomya* reaches with its siphons not to the surface, but into the prolonged burrows of the ghost shrimp, *Callianassa*. The basket shell, *Aloidis*, is bilaterally asymmetrical, with a larger right valve embracing the left. It has returned to shallow burrowing and fixation with a single byssus thread.

The main habit of the Myacea, that of deep and permanent burrowing, has been passed on to the superfamily Adesmacea (p. 59) (Fig. 20). This consists of the Pholadidae or piddocks (boring in soft rock or clay), and two families of wood-borers, the less-modified Xylophaginidae and the very specialized shipworms, Teredinidae. The ligament is much reduced in the Adesmacea, and the shell valves rock upon their hinge points in the transverse plane. The foot is used as an attachment disc whilst rotary boring movements are made. Evolution culminates in the extraordinary modifications of the genus *Teredo*. The naked siphon dominates the whole morphology of the animal; the visceral mass is small and anterior, the shell being used only as an abrading tool.

A very different habit of burrowing, with rapid vertical mobility in shifting substrata, is found in the Solenacea, or razor shells. Here the shell is long, light and razor-shaped, with the umbones and hinge at the anterior end, and the posterior territory of the shell enormously elongated. By contrast the siphons themselves are short. The foot is a long compressed plug that can be thrust downwards from the gape at the anterior end of the shell. The uniform cross-section of a razor shell well adapts it to sliding through sand, and its smoothness permits the great speed of burrowing that is the keynote of the razors and their allies. This design of bivalve has appeared separately several times. Douvillé was the first to separate the superficially similar *Ensis* from *Solen*, the former now being placed with the Heterodonta. Recently Owen has advocated a further cleavage by placing the East African estuarine, *Glaucomya*, in the Veneracea: so we now have a venerid razor shell as well as – in *Petricola pholadiformis* – an elongate venerid that has evolved parallel with the pholads. Finally, in *Psammosolen*, *Pharus* and *Solecurtus*, we have razor-shaped tellinids.

4   The last lamellibranchs
Orders Anomalodesmata and Septibranchia

These bivalves are apparently the latest evolved of the lamellibranch series. The relationships and the mode of life of the Anomalodesmata are still imperfectly understood and the group may be found to be polyphyletic: its members agree, however, in being all hermaphroditic and in having the outer demibranch reduced and upturned dorsally. From the frequent weakness or loss of the hinge-teeth and the fusion of the mantle margins, they may be suspected of having part-abandoned a former deep-burrowing habit. As in *Aloidis* – among the Myacea – the two shell valves may become unequal. Some species may lie recumbent on one side. In the Pandoridae and Myochamidae the valves show the asymmetrical *pleuroconch* condition, with a deeply convex right valve and a flatter left one. *Pandora* and *Myodora* may burrow shallowly by inserting the shell horizontally or obliquely into the sand. The fragile valves of the burrowing family Laternulidae (*Cochlodesma, Laternula* and *Periploma*) are also un-equal, the right being more convex. British *Cochlodesma praetenue* embeds in sand to a depth of up to 8 cm and lies horizontally on one or other side. Its siphons are long and separate, making mucous tubes as in Thraciidae; the inhalant one alone reaches the surface, the exhalant opening into a blind horizontal gallery. The Lyon-siidae show a progression towards sessile life: *Lyonsia* lives freely at the surface, *Entodesma* attaches by its byssus in crevices, and the American *Mytilimeria* nestles deeply in ascidian tests, becoming spherical and heteromyarian. The Thraciidae have returned to deep burrowing, and line their siphonal tubes with mucous secretion. The Australasian Chamostreidae have – like the Chamidae – become firmly cemented to the substrate by the deep right valve, the left forming a lid.

A strange offshoot is seen in the family Clavagellidae, the watering-pot shells, surely the least recognizable of all bivalves. As in *Teredo* the conjoined siphons are much the largest part of the animal and secrete a strong calcareous tube. *Clavagella*, which may reach nearly a metre in length and 5 cm in diameter, is embedded but cannot actively burrow. The valves are functionless and excessively small, the left one in *Clavagella* and both in *Brechites* being fused to one side of the tube. *Brechites* is of smaller size and burrows vertically in mud. Inhalant and exhalant currents both pass through the open posterior end. The anterior end, the place of the original pedal gape, is convex and perforated by small holes like the rose of a watering-pot. Of the evolution and adaptive significance of this family we yet know little.

The three genera *Cuspidaria, Cetoconcha* and *Poromya* are best given the status of a separate order, Septibranchia. They are evidently derived from surface-living Anomalodesmata. The septibranchs are little known at first hand, but are very beautifully specialized. As we have seen (p. 94), they are 'lamellibranchs without gills'. They replace these with a perforated muscular septum used for pumping water through the pallial cavity. In many ways the septibranchs are the most peculiar of the bivalves, retiring from a filter-feeding life to become scavengers or carnivores, ploughing through the plantless ooze at unlighted abyssal depths.[272, 318]

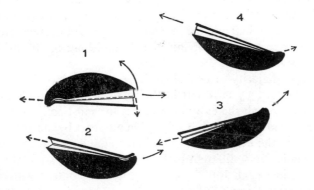

Figure 58   Pecten maximus
Swimming movements by water expulsion (*after Yonge*)
1 turning over
2, 3 swimming hinge forward
4 swimming hinge behind
broken line, water jet; entire line, direction of movement

11   The first and last cephalopods

This is the only class of molluscs as a whole pelagic. Its members are the largest, most active and most delicately specialized. Cephalopods are organized upon a more stereotyped pattern than gastropods or lamellibranchs; but this is a pattern of high success, for which advances in nervous co-ordination are supremely responsible. Still-mann Berry – speaking of the exquisite adjustment of modern cephalopods to their environment – cites their complex chromatophore systems, colour change and light organs; their interplaying systems of finely balanced musculature with few or no skeletal hard parts; the delicate balance between eye, sucker and chromatophore, or mantle arms and fins; and the innumerable types of hectocotylus and spermatophore, often involving most astonishing modifications in sexual behaviour.[66]

The great majority of cephalopods are now extinct, known only from the shells they have left in the rocks. The nautiloids ran un-challenged through the early Palaeozoic, the most active carnivores in the seas; and were then outstripped in numbers by the ammonoids which – with the belemnoids – continued abundant until the end of the Mesozoic. Very different methods are needed to study the living and the extinct cephalopods, and we must be content with a very partial understanding of the fossil forms. There is only one genus surviving that can tell us much about the way of life of Palaeozoic cephalopods: by good providence the living *Nautilus* has preserved a very ancient organization, and – after setting aside possible special-izations and peculiar features – we still have in the three recent *Nautilus* species living fossils of enormous value.

Nautilus, as we have seen, is very unlike all other living cephalo-pods. It has a large, many-chambered external shell, and its funnel is formed of two overlapping lobes. The tentacles are represented by numerous retractile filaments, with neither suckers nor hooks. The eye is of the simple pin-hole camera type, an open vesicle without cornea or lens. There is no ink-sac and no vivid play of chromatophores. There are two pairs of ctenidia and osphradia, as well as of auricles and kidneys. Most of these features in *Nautilus* are surely primitive; but it may not be safe to regard the modern survivors as typical in every way of the Palaeozoic cephalopods. In particular this may apply to the duplication of the palliopericardial organs. Rather than class all fossil ammonoids and nautiloids with the 'Tetrabranchiata' as was once the practice, we should remember that this condition may be an aberrant feature of *Nautilus* and its immediate relatives only. It could – on the other hand – have been widespread in the past, but this we shall probably never know. Flower has brought some evidence that the structure and large number of the tentacles of modern *Nautilus* may be a specialized character, not representative of most extinct forms.

The Nautiloidea and the Ammonoidea are each vast groups. The modern practice is to regard them as separate sub-classes and to rank the belemnoids and the modern cephalopods together in a third sub-class, the Coleoidea. Nautiloids first appear in the Upper Cambrian, ammonoids in the Upper Silurian. The former almost, and the latter entirely, die out at the end of the Cretaceous. A recent reckoning allows some 300 genera of nautiloids with about 2500 species. Flower and Kummel have grouped these into 75 families, forming 14 so-called 'orders'. The ammonoids were even more numerous, with 163 families and about 600 genera, divided into two main orders and six sub-orders. The largest members of the Ammonoidea are the Cretaceous giants *Pachydiscus septemarodensis*, like cartwheels two metres across. The greatest of the nautiloids must have been the straight forms of the genus *Endoceras*, some of which reached 4·5 m in length.[7]

The Nautiloidea possessed smooth, simply sculptured shells with gently curved septa attached to the wall of the shell by straight sutures. The Ammonoidea often bore heavy sculpture and in most genera the septa were wrinkled and deeply convoluted at the edges, forming intricate sutures, folded and refolded with incredible complexity. Such a septum – it has been suggested – served to increase the animal's surface of attachment in the body chamber of the shell; or it may have strengthened the shell against alterations in external pressure on changing depth. But of the real adaptive significance of ammonoid sutures we know almost nothing. The initial chamber of the nautiloid shell is a short one, situated opposite the blind end of the siphuncle; in ammonoids – as also in belemnoids and *Spirula* – it is spheroidal. There may have been fundamental differences between the ammonoid and nautiloid animals; but in many features the two groups have evolved on parallel lines, and we shall constantly meet with convergent adaptations in each.

Limited as our evidence may be, we can make some reconstruction of the life and habits of these mighty groups; and in certain interpretations the most cautious may agree.[65] The first nautiloids in the Upper Cambrian were no doubt also the first cephalopods. They had a cap-like, slightly curved shell, which increased in height as its apex first became occupied by gas-filled chambers. The siphuncle was widely open and some of the viscera probably extended into it. The foot may at first have retained its flat sole for creeping, but – as the distinctive cephalopod pattern emerged – it soon spread forward to surround the mouth and its edge became divided into a fringe of tentacles. As the growth of the chambers increased and the shell lightened, locomotion by water jets became possible. The animal could take short spurts upwards or backwards and dart nimbly off the ground (see p. 27).

Early nautiloids with short conical or cap-shaped shells may have swum upright. The first really fast pelagic forms were probably, however, the long, straight 'orthocones' of the *Orthoceras* type, moving horizontally like squids. Many of these species must have lurked close to the bottom, like modern sepioids, as we are sometimes able to deduce from the countershaded markings on the upper side of the shell alone, long and parallel stripes in the Ordovician *Geisnoceras*, zigzags in *Kionoceras*. A few early nautiloids such as *Gonioceras* (Ordovician) became strongly adapted to the bottom, with a flat wide-spreading shell and close-spaced septa, superficially resembling a cuttlebone.

The long orthocones varied from a few inches in length to 4·5 m in *Endoceras* (Ordovician).

Flower has described some shell trails and supposed tentacular impressions of the orthoconic nautiloid *Orthonybyoceras*; if his interpretations are correct, they give a fascinating insight into the habits of these creatures. The shell trails are straight and shallow, a little more than the length of the shell (see Figs. 60). Some are rounded behind and end abruptly in front; these – it is suggested – were made by the animal alighting after swimming horizontally. Other trails are rounded at both ends, as by the animal touching down briefly before swimming away again. Their arrangement makes it clear that the orthoconids could swim forward, as does a modern squid with the funnel pointed back, or could swing on their apex through a wide

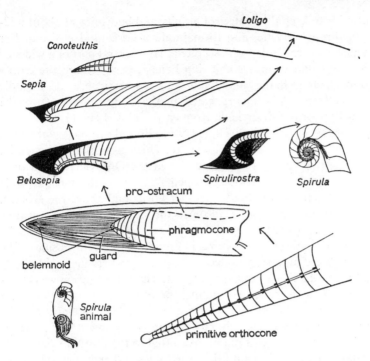

Figure 59 Derivation of the modern cephalopod shell (*based on Shrock and Twemhofel*)
(*inset, right*) modern *Spirula*

angle to change direction by steering with the funnel. Some shorter trails, as of the nautiloid *Fucoides graphica*, were radially clustered, suggesting congregation about a mass of food debris, or even a concourse of animals for mating as is found in the modern *Loligo*. In front of many trails of *Orthonybyoceras* was found a crescentic group of impressions attributed to short curved arms, apparently six to fourteen in number. These nautiloids may therefore have had a relatively small number of quite strong tentacles, which could be used for grasping the bottom or holding the animal steady.[122]

Even the longest of cephalopod shells must have been extremely buoyant, inconveniently so when resting near the bottom, since an apex filled with gas would tilt the head and body against the ground. Various expedients seem to have been employed by the straight nautiloids to keep the centres of gravity and buoyancy at about mid-length throughout life, and to maintain a horizontal position. For instance, calcareous deposits secreted in the ventral part of a widened siphuncle could form a stabilizing ballast, partly filling the empty chambers. Or, as in the curved Silurian Ascoceratidae, a dorsal saddle of air cells could be incorporated in the body chamber from the forward growth of diverticula from earlier closed chambers.

Coiling of the shell in a plane spiral allowed more elegant swimming and a freer power of manoeuvre. This was initiated very early, soon after the rise of the orthocones in the Ordovician; and once begun it seems to have been pushed further, as the most feasible method of holding the body chamber horizontal above the ground, rather than thrusting the head on to the substrate by the buoyancy of the apex. Forward migration of the centre of gravity could be retarded by coiling; and more rapid growth on the ventral side of the shell produced an openly coiled *gyrocone*. Even here increasing body weight would still lead to a forward tilt. The final solution was either a many-coiled *ophiocone* with numerous volutions all in contact, or a *nautilicone*, found in many Palaeozoic genera and in modern *Nautilus pompileus*. Here the later chambers completely invest the previous whorls, and the centre of gravity is shifted to the centre of the coil.[65]

Curved and straight nautiloids (*cyrticones* and *orthocones*) had run their course by the Triassic, and long before that – in the Devonian – had been outnumbered by coiled forms.

The first ammonoids evidently arose from coiled nautiloids in the later Silurian, the earlier representatives being the goniatites. They soon became much more varied than the nautiloids in the details of the siphuncle and sutures. The shell differed too in having the aperture closed by a form of operculum, either a single horny plate (*anaptychus*) or two calcareous plates (*aptychi*). The siphuncle generally lay 'ventrally', towards the lower side of the shell. Fig. 60 illustrates some types of sutures, beginning with the *goniatite* form, folded into simple *lobes* and *saddles*. These were succeeded by *ceratite* sutures (Permian to Triassic), with the saddles smooth and the lobes crenulated, and finally by the *ammonite* type, with both lobes and saddles intricately crenulated.

The advent of the ammonoids saved the Cephalopoda from decline: they ensured a continued abundance of coiled shells until the end of the Mesozoic, and they exploited many of the shell adaptations that the nautiloids had already begun. In lightening, streamlining and irregular coiling they went much further. Most nautiloids

– as indeed the modern *Nautilus*, though still a good swimmer – are blunt-prowed, and the shell is typically rounded in cross section. Ammonoids are often more compressed, often with angled edges, and sharp cut-water keels. Such keeled forms – as *Oxynoticeras*, *Oxycerites* and *Prionodoceras* – are referred to as *oxycones*, and were obviously the most active swimmers. There is an analogy among the small pelagic gastropods, with the rounded shell of the slow-swimming *Limacina* making leisured vertical migrations, and the keeled, compressed shells of *Oxygyrus* and *Atlanta* which are fast swimmers. The heavier shells with strongest sculpture probably belonged to ammonoids dwelling close to the bottom. All coiled species must, however, have been able to swim with the funnel in the manner of *Nautilus*. The ventral edge of the aperture has always a 'hyponomic sinus' or notch, presumably for the passage of a funnel.

Almost from the beginning of the nautiloids passively floating types developed. Thus in the Silurian we find plump vase-shaped *brevicones*, derived from orthocones, that must have drifted head downwards from the gas in the apical chambers. Their apertures are much contracted, often reduced to a mere slit, with a median exit for the funnel, and side notches for the eyes, which were perhaps borne on long stalks. Several pairs of arms probably issued by further side notches, of varying number, as in *Pentameroceras, Hexameroceras, Septameroceras* and many others. The coiled *Ophidioceras* and the horn-shaped *Phragmoceras* have also apertures like these. Much later, in the Jurassic, we find the coiled ammonoid *Morphoceras*, with only a narrow opening for the funnel, and separate side fenestrae for the arms and eyes. Such animals must have been unable to capture or ingest bulky food. They were almost certainly slow-moving, and were very possibly plankton-feeders. They perhaps had a thin widespread arm web for ciliary food collecting, or developed slender tentacles for deposit feeding. Such microphagous habits have been developed in several other carnivorous groups, as for example in anemones, jellyfish, starfish, pteropods and even basking sharks. Alternatively the narrow-mouthed cephalopods may have been predators with a suctorial proboscis, breaking down food externally as do many modern decapods by salivary enzymes.

In the latter half of the Mesozoic there appear many advanced ammonoids whose mode of life is even more difficult to reconstruct (Fig. 60). *Scaphites* begins as a coiled ophiocone, is then straight, and has a final recurved part where the adult lives. In *Hamites* the immature shell is also straight, and sharply recurves in the adult. In

Ptychoceras the recurved limb lies in contact with the previous straight part of the shell. *Heteroceras* has an open corkscrew spiral, the later part recurved, and *Spiroceras* has reverted to an open gyrocone. Such latter-day ammonoids were once regarded as 'degenerate' evolutionary lines, foretelling extinction. But it is surely unsafe to apply these quasi-moral judgments to populations still flourishing and numerous, and we may be pretty sure that the Jurassic and Cretaceous seas were populated by no handicapped or inefficient race of cephalopods. These ammonoids must have given up active swimming; and Trueman, by calculating the axis joining their centre of gravity and centre of buoyancy, has shown the posture in which many species floated. Like normally coiled ammonoids, they rested with the aperture raised above the bottom. Here they may have drifted almost passively, with the head conveniently tilted for catching sluggish prey. They could make leisured saunters, or up and down movements by the use of the funnel, and their buoyancy would increase when the animal was expanded, and be lowered when the soft parts lay compactly in the body chamber. The funnel must have given them active control of posture, and we need not consider such species in any sense as victims of an air-filled shell, mechanically restricted to any one attitude.

From early times helicoid nautiloids and ammonoids began to re-invade the bottom. Already in the Devonian the nautiloids had produced the flatly trochoid shell *Trochoceras*. In the Triassic appeared the long spiral-coiled ammonoid *Cochloceras*, and in the Jurassic came *Turrilites*, with a shell like a sinistrally coiled *Turritella*. Such shells may have been trailed on the substrate, or if air-filled and unballasted would have been light enough, though rather clumsy to carry aloft. The animals probably lived rather like octopods, crawling nimbly with the arms, and having the same facility for taking quick funnel spurts off the ground. A final aberrant state is the Japanese *Nipponites*, coiled in a complicated series of Us in varying planes. Whether or not it was sessile, it must have been immobile and possibly – like the irregular vermetid gastropods – trapped its food with mucus or cilia.

The phylogeny of fossil cephalopods must be approached as cautiously as their ecology. Spathe's review of the history of the Cephalopoda sharply challenges many of the older views, particularly the belief in 'programme evolution', with a grand evolutionary progress through straight, curved, open-coiled, tightly coiled and secondarily straight shells. The first nautiloids of all were in fact small, slightly curved cyrticones. Alongside the straight orthoceratids of the Ordovician and Silurian existed already a wealth of coiled shells of all graces. Straight shells, though earlier more abundant, co-existed with coiled until the end of the Palaeozoic; there was no successive replacement in time by cyrticones, gyrocones and nautilicones. Most open gyrocones and some cyrticones are in fact secondarily uncoiled nautilicones.[65] The first ammonoids were already coiled. As well as by forms like *Hamites* and *Scaphites*, they were overtaken in the Cretaceous by the secondarily straight ammonoids of the Baculitidae, extremely similar to early orthoconic nautiloids. The ammonoids have left no descendants. The nautiloids, though they dwindled earlier, have persisted longer, having a short renaissance in the Tertiary with coiled genera such as *Aturia*. *Nautilus* itself remains living. The last of the straight orthoceratids gave rise as well to a vigorous straight-shelled line that expanded in the Mesozoic. These were the belemnoids, from which modern cephalopods have sprung; and it is to their descendants that we must now turn in our survey of the sub-class Coleoidea.

The Belemnoidea and modern Decapoda

The Coleoidea (Figs. 22, 25) fall naturally into two main orders: the first and larger, the Decapoda, with ten arms, body fins and an internal shell or a vestige of one, and the second, the Octopoda, with eight arms of equal length, no shell and usually no fins.* The modern coleoids have fewer species than most other molluscan sub-classes. They are not, however, a dwindled relict group, but rather a new expansion of the Cephalopoda into molluscs of superior power and activity to all others. Metabolic rate, sensory and locomotor powers have vastly improved, and the average body weight has increased a hundredfold. Only a few genera live in restricted habitats on the bottom, and it is this greater mobility and wider range that explains their relative lack of speciation. Ecologically as well as physiologically we are dealing with molluscs at a new level of independence.

The Belemnoidea (Fig. 60) – the first sub-order of the Decapoda – arose in the Triassic, and disappeared, apart from one Tertiary family (Neobelemnitidae), in the Cretaceous. They are distinguished from their straight nautiloid ancestors by the shell being internal. As we have already seen in Chapter 3, the belemnoid shell had three

*See Vampyromorpha, p. 229.

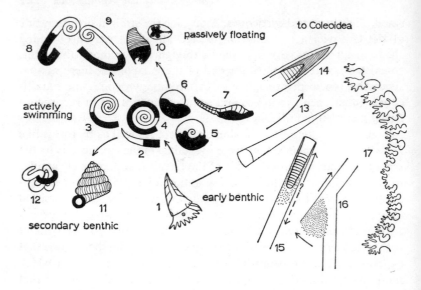

Figure 60 Evolution of the coiled cephalopod shell
1–12 coiled shells in their natural postures
  1 an early brevicone
  2 cyrticone
  3 gyrocone
  4 ophiocone
  5 ammoniticone
  6 nautilicone
  7 *Ascoceras*
  8 *Macroscaphites*
  9 *Lytoceras*
  10 *Hexameroceras*, also shown in apertural view
  11 *Trochoceras*
  12 *Nipponites*
  13 primitive orthocone
  14 belemnoid

15–16 trails of *Orthonybyoceras* (*after Flower*)
  15 straight trail with shell *in situ*
  16 trail showing change of direction

17 an ammonitid suture

parts: a small chambered *phragmocone* corresponding to the whole nautiloid shell, a solidly calcified *rostrum* which is the cigar-shaped part usually found fossil, and, extending forward, a broad shield, the *pro-ostracum*. The belemnoids were evidently rapid funnel-swimmers like a modern *Loligo*, the rostrum giving them an arrow-like rigidity, the phragmocone giving buoyancy at the centre of the body, and the pro-ostracum providing attachment for the mantle. The external mantle would allow improved jet-locomotion, and its tough muscles have in a few cases made possible the preservation of impressions of soft parts that we never obtain from the delicate tissues of ammonoids and nautiloids. Traces survive of triangular fins as in *Loligo*, of the head with rather feeble jaws, of the funnel and of the ink-sac. Most interesting, the arms were provided with one or two rows of hooks instead of horny suckers. Whether all belemnoids were hooked is not known, nor is there general agreement on the number and relative length of the arms. Some authors figure six, others eight, apparently of equal length. Later forms at least appear to have had ten, and so – in the strict sense – can be admitted as 'decapods'.

Belemnoids were generally of modest size, the animal probably about 30 cm in total length. The largest probably reached well over 2 metres. There were numerous species, grouped in a recent classification into sixty-three genera and five families. They seem to have shown rather little adaptive radiation, though genera like *Hibolites* with a light spear-shaped rostrum must have contained fast swimmers. Others – such as *Belmnites giganteus*, with a heavy conical rostrum – were slower and benthic. Belemnoids are thought to have fed on fish, crustaceans and at times on each other. From the finding of 'graveyard' formations of fossils, they must at times have swum together in immense swarms.

Living decapods are classed in two sub-orders, the Teuthoidea and the Sepioidea.* In all of them, as we have seen, the shell is reduced in importance. The main existing types are the open-coiled phragmocone of *Spirula*; the 'shell' of *Sepia*, with the upper side of the phragmocone wide and flat and with a small rostrum; and the 'pen' of *Loligo*, uncalcified and with a horny gladius and shaft representing the pro-ostracum. Various Tertiary fossils provide annectant forms leading down to these modern survivors (Fig. 59). *Spirulirostra*, with a curved phragmocone and persistent rostrum, foreshadows *Spirula*;

*For modern classification, see Donovan (1964).[107]

Belosepia, with the lower side of the phragmocone reduced, leads to *Sepia*; and from *Beloteuthis* and *Palaeololigo*, with large pro-ostraca, was produced the pen of the squids.

The Sepioidea, together with the addition of the Loliginidae from the Teuthoidea, correspond to the formerly recognized group Myopsida, distinguished from other families (Oigopsida) by their closed external cornea. The sepioid families Sepiidae, Sepiadariidae and Sepiolidae are on the whole coastal in distribution, adapted to benthic life and of comparatively small size. The little *Spirula* – 7·5–10 cm in length – is probably a representative of an isolated bathypelagic group of sepioids. Bruun, who has studied it alive, found it to float or swim actively in a slightly oblique position with its head downward. It is kept upright by the small gas-filled shell, surrounded by soft tissues.

Both the sepioids and the octopods show more speciation than the oceanic Teuthoidea, but the latter easily predominate in numbers of families and genera. Only the coastal Loliginidae show many species. The loliginids and sepioids are certainly much better known than the squids of deeper water. The majority of the Teuthoidea can indeed have been seen alive only by fishermen. Some of the deeper-water Teuthoidea are giants of the seas; others – as the Cranchiidae – are relatively small. We shall mention in turn some of the families of greatest interest.

The mightiest of the cephalopods are the several species of the Architeuthidae, with such proud names as *Architeuthis princeps* and *Architeuthis dux*. These may exceed 15 metres in total body length. Three-quarters of this is accounted for by the tentacles, the shorter arms reaching up to 4·5 m. The fins, though large, are not excessively developed, and the locking apparatus of the mantle (p. 65) is feeble. This suggests that the architeuthids are not rapid swimmers; they may be taken off the continental slope in depths of only 180–360 m (100–200 fathoms) where they seem to feed on the larger benthic animals. They are also known as larvae, being the small squids once called *Rhynchoteuthis*, with the two tentacles joined in the form of a spout.

Much more accomplished swimmers are the medium giants of the family Sthenoteuthidae. *Sthenoteuthis caroli*, occasionally entering the northern North Sea from the Atlantic, may reach up to 90 cm in body length and 2 metres overall. It is exceedingly active, with a strong locking apparatus and broad rhomboidal fin vane and can shoot from the water like a rocket, sometimes landing on shipboard. Hardy remarks that this species is 'very adept at summing up a situation and quickly taking the only way out . . . it thus skilfully avoids traps and nets'. A smaller related species, *Ommastrephes sagittatus*, is sometimes stranded in large numbers on British coasts. The most adept 'flying squids' we have already mentioned (p. 67) as belonging to the family of hooked squids, Onycoteuthidae, using their fin vane and funnel for shooting from the water and gliding like flying fish. Even larger vanes are found in the deep-water Octopodoteuthidae: in species of *Cucioteuthis*, the eight arms and two tentacles are all short and the two fins together form a circular expanse many times larger than the rest of the body. In the family Thysanoteuthidae the tentacular arms have two rows of filaments, as well as suckers, running their whole length. Two triangular fins are attached the whole length of the body, giving it a rhomboidal shape like a ray.

Still greater specializations are found in the bathypelagic family Chiroteuthidae (Fig. 25). These small squids are thin-bodied and cylindrical, specializing, as the names *Chiroteuthis* and *Mastigoteuthis* suggest, in extraordinarily long, whip-like tentacular arms. These are non-retractile and are equipped with suckers spaced all the way along; they may reach six times the body length, or – in *Chiroteuthis veranyi* – some 60 cm (24 ins.). Of the short arms, the ventral two are very flat and wide. The fin vane lies right beind the body, forming a flat, circular expanse; in *Grimalditeuthis* there is a double vane, from a pair of semi-circular fins at either side. *Doratopsis vermicularis*, with a terminal circular vane, is the slenderest of all cephalopods, with a flexible body the thickness of a pencil.

Some weirdly adapted squids belong to the two deep-water families Histioteuthidae and Cranchiidae. The first are of smallish size, less than 30 cm (1 ft) overall, sometimes much smaller, and studded with light organs. In *Calliteuthis* the eyes are asymmetric, the right one smaller and sunken with a circlet of light organs which the left one lacks. *Histioteuthis* species are reddish-purple to black and resemble small vampire squids (see p. 229) with the six dorsal arms united by a membranous web, whose medusoid movements virtually replace the funnel in swimming as in the cirroteuthoid octopods.

In the Cranchiidae we have about twenty genera of small squids, on the whole planktonic within about 100 metres of the surface, though some go much deeper. They present some odd shapes and

H

many of the described species are probably larval. *Cranchia* itself has a plump, vase-shaped body, narrowest at the mantle edge, with two long tentacles and the other arms very short. In *Bathothauma*, a deepwater genus, these short arms are carried up in a small rosette on a long penduncle which bears the mouth. In most cranchiids the posterior fins are small and rounded, but *Galiteuthis* – which is a fast swimmer – has a slender body about 20 cm in length, tipped with a large triangular fin vane. The eyes of cranchiids are always prominent and sometimes mounted on stalks. In the small *Sandalops melancholicus* the stalks are bent down at right angles, and the eyes look ahead with an unmistakable air of sadness which has given rise to the specific name.

Many cephalopods show a striking bioluminescence.[66] This is especially a property of the oigopsids, found in three-quarters of the known species, particularly those living in deep waters. Relatively few myopsids are luminescent, and in the Octopoda only two deepwater species – *Melanoteuthis luceus* and *Eledonella alberti*. In four families of squids every species possesses light organs, namely the Lycoteuthidae, Lampadoteuthidae, Bathyteuthidae and Enoploteuthidae, as well as most if not all of the Cranchiidae and Histioteuthidae. In these families light is produced by the photogenic tissues of the animal itself, and special light organs may be distributed in almost every part of the body. In the cranchiids they occur – as well as elsewhere – on the lower surface of the eyeballs, where they can be covered by movable folds of skin. Other families may have optic light organs without lids. They are generally found widely on the integument of the mantle, head and arms, especially in circlets around the eyes. In luminescent Chiroteuthidae they extend right down the long tentacles. There may also be intrapallial light organs on the visceral mass. These rely for the effect on the transparency of the tissues; and the animal must resemble the apocalyptic beast, 'full of eyes within'. The histology of the light organs has had much attention from authors such as Joubin, Hoyle, Chun and Stillmann Berry.[66] They become extraordinarily complicated and diverse, especially in the Enoploteuthidae and Histioteuthidae. There is usually a nest of primary photogenic tissue, to which may be added 'reflector mechanisms, pigment cups, lenses, diaphragms, directive muscles, windows, colour screens and accessory photophores'. For descriptions of the light produced, the old expedition reports are full of interest. To quote from Hoyle's translation of Chun's Valdivia Report on *Lycoteuthis diadema*:

Among all the marvels of colouration which the animals of the deep sea exhibited to us, none can be even distantly compared with the hues of these organs. One would think that the body was adorned with a diadem of brilliant gems. The middle organs of the eyes shone with ultramarine blue, the lateral one with a pearly sheen. Those towards the front of the lower surface of the body gave out a ruby red light, while those behind were snow white or pearly, except the median one which was sky blue. It was indeed a glorious spectacle.

Then Watase's description of the Japanese inshore squid *Watasenia scintillans*, the *hotaru-ika*, fortunately common at times in shallow water:

When the animal is about to produce light the chromatophores covering the spots will concentrate and remove themselves, thus opening a way for the light. The light is so brilliant that it seems like a sunbeam shot through a tiny hole in a window curtain.

Luminescent myopsids employ a different means of lighting by an association between the animal and symbiotic luminescent bacteria. In Loliginidae and Sepiidae – as well as luminescent bacteria harboured on the skin – the ducts of the accessory nidamental glands are filled with masses of photogenic bacilli or cocci. Light may be produced by internal illumination, or emission of the contents into the water. The bacteria are transmitted to the next generation intimately attached to the egg membrane. In the Sepiolidae (*Sepiola* and *Rondeletia*) there are more complex light organs, formed by specialization of part of the accessory nidamental gland, and the provision of a pigment sheath, reflector and lens. In the deep-water sepiolid *Heteroteuthis dispar* (2150–2700 m (1200–1500 fathoms)) the light is not finally proved to be due to bacteria; instead of a cloud of ink, an attacking fish receives 'a veritable bombardment of liquid light' in greenish clouds of faint cobalt patches.

What is the significance of bioluminescence? A. C. Hardy gives an interesting discussion of light production in deep-water crustacea, and finds it associated with migrations over a long depth range rather than with a permanently deep habitat. This seems true of cephalopods as well – the permanently abyssal octopods, for example, do not produce light at all. As well as repelling predators, light may be of direct use to the animal itself, though whether – as has been suggested – it is especially employed in courtship display, for hunting food, or for recognition of fellows, we do not know. As we have suggested before, cephalopod life is so fundamentally centred round the visual sense that it may have been difficult – at all events for deep-migrating oigopsids – to substitute any other sense in darker waters. Perhaps only by taking with them their own light supply have such essentially visual animals as squids been able to penetrate the unlighted depths.

The Octopoda

The octopods have lost all trace of the shell, except for a last vestige of it in the fossil *Palaeoctopus*. The two tentacular arms have disappeared and the other eight arms themselves form long tentacles, arranged in a circlet linked by a web around the mouth. *Octopus* and *Eledone*, with rounded body, no fins and leisured swimming, are very typical of inshore Octopoda. They have much more contact with the bottom than most decapods. The eight arms no longer merely hold the captured prey. They themselves are organs of attack, and are exploratory and tactile members as well. The Octopoda have a finely developed sense of touch, and those that live at the bottom employ colour change as well, though never with such subtlety as in the cuttlefish. We must not regard all Octopoda as like *Octopus* in habits. There are three other very distinctive modes of life shown by offshore genera – the surface-dwelling, the bathypelagic and the permanent abyssal-benthic.

The Argonautacea are the typical octopods of the upper waters of the open sea. They are found from the surface to 900 m (500 fathoms), being proficient but not fast jet-swimmers. Above all other cephalopods they exhibit sex dimorphism. In *Argonauta* (Fig. 25) itself the sex difference is greatest, with two of the arms in the female modified to secrete a graceful papery 'shell'. This contains the bunches of eggs, which are small and extremely numerous. The males are dwarfed with an elaborate and often autonomous hectocotylus arm. In *Ocythoe*, a pelagic form without a 'shell', the dwarf male is tiny enough to shelter in empty salp tests like the amphipod *Phronima*.

In very deep waters, but not usually living at the bottom, occur the bathypelagic octopods such as *Eledonella, Vitreledonella* and *Amphitretus*. These have a long depth range and no particular modifications for benthic life. The greatest depth record of any cephalopod is that of *Eledonella*, reaching 2949 fathoms (*c.* 5300 m). Another genus, *Japetella,* is interesting in having a pelagic larva, supported at the surface by a cuticular test produced into irregular filaments. *Amphi-*

tretus, which has the arms half-webbed, possesses 'telescopic' eyes raised on stout peduncles from the dorsal side. The whole body is clad in a loose sheath of transparent jelly.

The deep benthic octopods of the superfamily Cirroteuthacea are the most strangely modified of all. They are web-swimmers, a logical development of the octopod pattern of eight radial arms joined by a basal web. In *Cirroteuthis*, *Cirrothauma* and *Opisthoteuthis*, as we have seen, the web is very large, reaching almost to the arm tips. Within the web the arm suckers are modified as a double row of filaments, used as tactile organs and for collecting fine particulate food. *Cirroteuthis* and *Cirrothauma* have a high conical body and swim slowly by opening and closing the web. *Opisthoteuthis* is flattened, rather like a webbed starfish, with mouth below, bulging eyes, small funnel and a pair of fins above (Fig. 25).

The cirroteuthids and many of the bathypelagic octopods have lost most of their firmness and muscular power. They develop a thick coat of jelly-like subcutaneous connective tissue, and the muscles themselves are sometimes degenerate and invaded by this jelly. *Cirrothauma* is as fragile in texture as a jellyfish, and one can read newsprint through its transparent body. With the slower web-swimming there is reduced muscular effort and no doubt a lower metabolic rate. In the Cirroteuthidae the branchial hearts are consequently very small, and the structure of the gill is simplified. The pallial aperture and the funnel are likewise very reduced. Feeding is microphagous, and though the jaws are retained, the radula – never of major importance in cephalopods – is quite lost in *Cirrothauma*. In total darkness the chromatophore system and the ink-sac are valueless, and they seldom survive. Light organs are unusual in any octopods. Finally – strange culmination for a cephalopod! – *Cirrothauma* has almost lost its eyes as well. In *Opisthoteuthis*, of the shallower continental shelf, these are still well developed.

With the webbed octopods were once classed the small deep-water vampire squids (*Vampyroteuthis*) for which – since the work of Pickford (1946) – we must recognize a special third order of living Coleoidea, the Vampyromorpha. The animal is deep purplish-black in colour, the female about 20 cm long and the male a little smaller. They swim by an arm web, beneath which the tentacles bear short cirri. The conical body bears large terminal fins and is covered with light organs, the one at the base of either fin having a large 'eyelid'. There appear at first to be eight equal arms, but close inspection reveals a unique character in two small retractile filaments like coiled tendrils, homologous not with the long tentacles of decapods but apparently with the second dorsal pair of arms.[255]

Cephalopod overview

We have spoken much of cephalopod efficiency and success, but M. J. Wells has given an interesting discussion of their physiological handicaps and relative failure. First, haemocyanin is an inefficient respiratory pigment. It never occurs in corpuscles and though the pigment is, in cephalopods, several times more concentrated than in other molluscs or in crustaceans, the oxygen capacity of the blood is still only 3–4·5 volumes per cent, compared, for example, with 10–20 volumes in a fish. In spite of such elaborate circulation boosters as auxiliary hearts and contractile veins, the returning venous blood is almost devoid of dissociable oxygen. Cephalopods quickly die under poor circulation and they may well have been limited in their ecological range by failure to find a more efficient oxygen carrier. The molluscan kidneys, though adequate in slower animals, are poor improvisations for cephalopods, and nitrogenous excretion is inefficient, taking place through glandular tissue around the afferent branchial vein, and also through the 'hepatopancreas' emptying into the gut caecum. The product is mainly ammonia, but includes some uric acid. But a blood pressure of only 5 mm of mercury would not filter the large volumes required for osmoregulation in a more dilute medium.[22]

In one of the most stimulating reviews of the cephalopods ever written, Packard[252] has surveyed their past and present interactions with bony fish, and the main changes achieved in converting an ancient cephalopod into a modern one. He compellingly argues that 'the vertebrates produced the coleoids'. Their convergences of structure and functional achievement came about not merely by the similar demands of the environment, but by their specific response to a vertebrate-dominated environment. Adaptation to the physical features has been less important than adaptation to the behaviour taking place within the environment, particularly the complex of behaviour that links predators and their prey. The vertebrates are thus not only the highest and dominant organisms in the food chain, but also the main source of selection pressures within behaviour-space.

The cephalopods achieved their breakthrough by jet propulsion and buoyancy, based on the chambered shell. Jet propulsion was not size-limiting, being more effective at moving large bodies than small: in combination with neutral buoyancy it permitted the evolution of the highest animals in the sea at that time. The ability of both cephalopods and vertebrates to move freely in the overlying water was like the development of flight by terrestrial animals. With independence of the ground, came the freedom to explore the whole possibilities of a new environment.

Through the Palaeozoic and most of the Mesozoic, the main advantage of the cephalopods over marine vertebrates was their vertical mobility, with buoyancy control by an osmotic pump, primed by the siphuncular epithelium. When the vertebrates invaded the adaptive zone of the cephalopods, vertical mobility was exploited to the full, and led the cephalopods to go deeper, to occupy a peripheral place in the seas, in short – *to diverge*. But it was the alternate tendency, to *converge*, with the adaptive zone moving in the vertebrate direction, that ultimately led to the coleoids' modern survival and success.

Packard has listed the convergences that were progressively able to be made, by cephalopods, occupying the major adaptive zone of the teleosts. These range through body form and locomotion, hydrostatic control, feeding and prey capture, and new organization of the central nervous system, sense organs and behaviour to subserve higher metabolic levels. We have already noted the striking convergence of the fish and cephalopod eye and vision, with parallels in the photographic system, colour change, cryptic patterns and behaviour.

'The evolution of fish and cephalopods' (writes Packard) 'is linked dynamically by the fact that, whenever an adaptive subzone is occupied by the two, an adaptive change in one will increase the selection pressure for a similar or alternative adaptive change in the other; for example, night-hunting fish acquiring physiologically effective quantities of visual pigment which enable them to see better in the dark will impose selection pressures for improved night vision on the pelagic squid population with which they are in competition. Thus parallelism and convergence between fish and cephalopods tend to continue or increase.'

Classification of the phylum Mollusca

Subphylum I   ACULIFERA

Class   APLACOPHORA
Subclass   *VENTROPLICIDA*
Families   Lepidomeniidae, Neomeniidae, Proneomeniidae,
   Parameniidae

Subclass   *CAUDOFOVEATA*
Family   Chaetodermatidae

Class   POLYPLACOPHORA
Order   Lepidopleurida
Family   Lepidopleuridae

Order   Chitonida
Families   Lepidochitonidae, Mopaliidae, Cryptoplacidae,
   Ischnochitonidae, Chitonidae

Subphylum II   CONCHIFERA

Class   MONOPLACOPHORA
Subclass   *CYCLOMIA*
Subclass   *TERGOMYA* (living) Neopilinidae

Class   GASTROPODA
Subclass   *PROSOBRANCHIA*
Order   Archaeogastropoda

Superfamilies	*Families*
Zeugobranchia	Pleurotomariidae, Haliotidae, Scissurellidae, Fissurellidae
Trochacea	Trochidae, Turbinidae
Patellacea	Acmeidae, Patellidae
Neritacea	Neritidae, Helicinidae

Order   Mesogastropoda

Archaetaenioglossa	Cyclophoridae, Pilidae, Viviparidae
Valvatacea	Valvatidae
Littorinacea	Littorinidae, Pomatiasidae, Acmidae
Rissoacea	Rissoidae, Hydrobiidae, Assimineidae
Cerithiacea	Cerithiidae, Potamididae, Cerithiopsidae, Melaniidae, Turritellidae, Siliquariidae, Vermetidae
Ptenoglossa	Ianthinidae, Scaelidae (Epitoniidae)
Aglossa	Aclididae, Stiliferidae, Eulimidae, Entoconchidae
Calyptraeacea	Calyptraeidae, Capulidae, Xenophoridae
Strombacea	Aporrhaidae, Struthiolariidae, Strombidae
Heteropoda	Atlantidae, Carinariidae, Pterotracheidae

Superfamilies	Families
Naticacea	Naticidae
Cypraeacea	Cypraeidae, Lamellariidae
Doliacea	Cassididae, Cymatiidae, Bursidae, Doliidae, Pirulidae

Order Neogastropoda

Buccinacea	Buccinidae, Nassidae, Galeodidae, Fasciolariidae
Muricacea	Muricidae, Thaididae, Coralliophilidae, Magilidae
Volutacea	Volutidae, Harpidae, Mitridae, Olividae, Marginellidae
Toxoglossa	Conidae, Turridae, Terebridae

Subclass OPISTHOBRANCHIA
Order Cephalaspidea

Bullariacea	Actaeonidae, Bullariidae, Retusidae
Philinacea	Scaphandridae, Philinidae, Atyidae, Runcinidae, Gastropteridae
*Pyramidellidae	

Order Anaspidea — Aplysiidae, Akeratidae
Order Thecosomata — Limacinidae, Cavoliniidae, Cymbuliidae, Peraclidae
Order Gymnosomata — Pneumodermatidae, Cliopsidae, Clionidae

Order Sacoglossa

Oxynoacea	Arthessidae, Oxynoidae
Elysiacea	Hermaeidae, Elysiidae, Limapontiidae

Order Acochlidiacea — Acochlidiidae
Order Notaspidea

Umbraculacea	Umbraculidae
Pleurobranchacea	Pleurobranchidae

Order Nudibranchia

Dendronotacea	Tritoniidae, Iduliidae, Lomanotidae, Dendronotidae, Scyllaeidae, Phyllirhoidae, Tethyidae
Aeolidiacea	Coryphellidae, Aeolidiidae, Facelinidae, Calmidae, Glaucidae
Doridacea	Dorididae, Bathydorididae, Dendrodoridae, Notodorididae, Polyceridae, Onchidorididae, Goniodorididae
Arminacea	Arminidae

*Provisionally placed here by Fretter and Graham (1949).

I

Subclass	*PULMONATA*	
Order	Basommatophora	
	Superfamilies	*Families*
	Actophila	Ellobiidae, Otinidae, Chilinididae
	Amphibolacea	Amphibolidae
	Patelliformia	Siphonariidae
	Hygrophila	Lymnaeidae, Physidae, Planorbidae, Ancylidae
Order	Stylommatophora	
	Onchidiacea	Onchidiidae
	Soleolifera	Vaginulidae
	Succineacea	Succineidae
	Tracheopulmonata	Athoracophoridae
	Achatinellacea	Achatinellidae
	Vertiginacea	Cochlicopidae, Vertiginidae, Enidae, Valloniidae, Clausiliidae
	Achatinacea	Achatinidae, Ferussaciidae, Subulinidae
	Oleacinacea	Testacellidae
	Endodontacea	Endodontidae, Arionidae
	Zonitacea	Zonitidae, Limacidae, Vitrinidae, Polygyridae
	Acavacea	Acavidae
	Bulimalacea	Bulimulidae, Urocoptidae
	Helicacea	Helicidae, Pleurdontidae, Fructicicolidae
	Streptaxacea	Streptacidae, Paryphantidae
Class	BIVALVIA	
Subclass	*PROTOBRANCHIA*	
Order	Nuculacea	Nuculidae, Malletiidae
Order	Solenomyacea	Solenomyidae
Subclass	*LAMELLIBRANCHIA*	
Order	Arcoida	
	Arcacea	Arcidae, Glycymeridae
Order	Mytiloida	
	Mytilacea	Mytilidae
Order	Pterioida	
	Pteriacea	Pteriidae, Vulsellidae
	Pectinacea	Pectinidae
	Anomiacea	Anomiidae
	Ostreacea	Ostreidae
	Pinnacea	Pinnidae
	Limacea	Limidae
	Plicatulacea	Plicatulidae
Order	Schizodonta	
	Trigoniacea	Trigoniidae
	Unionacea	Unionidae, Mutelidae, Aetheriidae
Order	Heterodonta	
	Astartacea	Astartidae

Superfamilies	Families
Carditacea	Carditidae
Sphaeriacea	Sphaeriidae, Coriculidae
Isocardiacea	Isocardiidae
Cyprinacea	Cyprinidae
Cyamiacea	Cyamiidae
Gaimardiacea	Gaimardiidae
Dreissenacea	Dreissenidae
Lucinacea	Lucinidae
Erycinacea	Erycinidae, Galeommatidae
Chamacea	Chamidae
Cardiacea	Cardiidae, Tridacnidae
Veneracea	Veneridae, Petricolidae
Mactracea	Mactridae, Amphidesmatidae
Tellinacea	Donacidae, Asaphidae, Semelidae, Tellinidae

Order Adapedonta

Solenacea	Solenidae, Glaucomyidae
Myacea	Aloididae, Myidae
Saxicavacea	Saxicavidae
Adesmacea	Pholadidae, Xylophaginidae, Teredinidae

Order Anomalodesmata

Pandoracea	Lyonsiidae, Pandoridae, Myochamidae, Chamostreidae, Thraciidae, Laternulidae, Verticordiidae
Clavagellacea	Clavagellidae

Order Septibranchia

Poromyacea	Poromyidae, Cuspidariidae

Class CEPHALOPODA

I. Subclass *NAUTILOIDEA*
14 'orders', 75 families
(Nautilidae living)

II. Subclass *AMMONOIDEA*
2 orders, 163 families (none living)

III. Subclass *COLEOIDEA**
Order Decapoda
Sub-order Belemnoidea (5 families, all extinct)

Sub-order Sepioidea

Spirulacea	Spirulidae
Sepiacea	Sepiidae, Sepiadariidae, Sepiolidae Idiosepiidae

Sub-order Teuthoidea

Loliginacea	Loliginidae
Architeuthacea	Lycoteuthidae, Enoploteuthidae, Abraliidae,

*Donovan[107] has now proposed a definite subdivision of the class Coleoidea into a number of smaller natural groups.

Superfamilies	Families
Architeuthacea (*cont.*)	Octopodoteuthidae, Onycoteuthidae, Gonatidae, Architeuthidae, Histioteuthidae, Bathyteuthidae, Ommatostrephidae, Sthenoteuthidae, Thysanoteuthidae, Chiroteuthidae, Cranchiidae Lampadoteuthidae
Order Vampyromorpha	Vampyroteuthidae
Order Octopoda	
Cirroteuthacea	Cirroteuthidae, Opisthoteuthidae
Bolitaenacea	Bolitaenidae, Amphitretidae, Vitreledonellidae
Octopodacea	Octopodidae
Argonautacea	Alloposidae, Tremoctopodidae, Ocythoidae, Argonautidae

Bibliography

Principal malacological periodicals

Journal de Conchyliologie (1850 . . . France)

Archiv fur Molluskenkunde (1868 . . . W. Germany)

The Journal of Conchology (1874 . . . U.K.)

The Nautilus (1886 . . . U.S.A.)

Proceedings of the Malacological Society of London (1893 . . . U.K.) now Journal of Molluscan Studies

Venus (1928 . . . Japan)

Journal of the Malacological Society of Australia (1957 . . . Australia)

The Veliger (1958 . . . U.S.A.)

Malacologia (1962 . . . U.S.A.)

Malacological Review (1968 . . . U.S.A.)

General reference books

1. FISCHER-PIETTE, H., and FRANC, A. (ed.), 1960. *Traité de Zoologie*, tome 5, fasc. 4 (Mollusques: Introd., Amphineura, Monoplacophora, Bivalvia (part)). Paris.
2. FRETTER, V., and GRAHAM, A., 1962. *British Prosobranch Molluscs*. London.
3. FRETTER, V., 1968. *Studies in the Structure, Physiology and Ecology of Molluscs*, Symp. Zool. Soc. Lond. 22.
4. FRETTER, V., and PEAKE, J., 1975. *Pulmonates*. Vol. 1, Functional Anatomy and Physiology. London.

5. HOFFMANN, H., 1938. 'Opisthobranchia', in *Bronn's Tierreich, III.*
6. HYMAN, L. H., 1967. *The Invertebrates: Mollusca I.* N.Y.: McGraw-Hill.
7. MOORE, R. C. (ed.), 1957. *Treatise on invertebrate palaeontology*, pt. I. New York (Ammonoids).
8. MOORE, R. C. (ed.), 1960. *Treatise on invertebrate palaeontology*, pt. I. Mollusca I. (Introd., chitons, gastropods, scaphopods, etc.).
9. NIXON, Marion, and MESSENGER, B. J. (ed.), 1977. *The Biology of Cephalopods.* London.
10. PRUVOT-FOL, Alice, 1954. 'Opisthobranchia', in *La faune de France.* Paris.
11. THIELE, J., 1931–1935. Handbuch der systematischen Weichtierkunde, 4 vols. Jena.
12. WILBUR, K., and YONGE, C. M. (ed.), 1964. *Physiology of Mollusca.* New York; Vol. 2, 1966.
13. YONGE, C. M. and THOMPSON, T. E., 1976. *Living Marine Molluscs.* London.

Reference books on special topics

14. BULLOCK, T. H., and HORRIDGE, G. A., 1965. *The Nervous Systems of the Invertebrates.* San Francisco.
15. CROFTS, D. R., 1929. *L.M.B.C. Memoir*, 29 Haliotis. Liverpool.
16. DAKIN, W. J., 1912. *L.M.B.C. Memoir*, 20 Buccinum. Liverpool.
17. EALES, N. B., 1921. *L.M.B.C. Memoir*, 24 Aplysia. Liverpool.
18. PURCHON, R. D., 1968. *The Biology of Molluscs.* London: Pergamon.
19. RUNHAM, N. W. and HUNTER, P. J., 1970. *Terrestrial Slugs.* London: Hutchinson University Library.
20. THORSON, G., 1946. *Reproduction and larval development of Danish marine bottom invertebrates.* Medd. Comm. Danm. Fiskeri- og Havunders. ser. Plankton 4(1). Copenhagen.
21. TOMPSETT, D. H., 1939. *L.M.B.C. Memoir*, 52 (*Sepia*). Liverpool.
22. WELLS, M. J., 1962. *Brain and behaviour in cephalopods.* London.

23. YONGE, C. M., 1949. *The Sea Shore.* London.
24. YONGE, C. M., 1950. *The Oyster.* London.
25. YOUNG, J. Z., 1966. *The memory system of the brain.* Univ. Calif.

Regional books on molluscs and shells

26. ABBOTT, R. T., 1955. *American sea shells.* Princeton.
27. ABBOTT, R. T. (ed.), 1959, *Indo-Pacific Mollusca.* Philadelphia.
28. ABBOTT, R. T., 1962. *Sea shells of the world.* New York.
29. ALDER, J., and HANCOCK, A., 1845–1855. *A monograph of the British Nudibranchiate Mollusca.* London.
30. ALLAN, J., 1960. *Australian sea shells.* Melbourne.
31. CZERNAHORSKY, W. T., 1967, 1972. *Marine Shells of the Pacific.* Vols. 1, 2. Sydney.
32. ELLIS, A. E., 1926. *British Snails.* London.
33. FORBES, E., and HANLEY, S., 1853. *A history of British molluscs and their shells.* London.
34. JEFFREYS, J. G., 1862–1896. *British conchology.* London.
35. KEEN, A. M., 1958. *Sea shells of tropical West America.* Stanford.
36. KIRA, T., 1959. *Japanese shells in colour.* Osaka.
37. KIRA, T., 1962. *Shells of the Western Pacific in colour.* Osaka.
38. POWELL, A. W. B., 1962. *Shells of New Zealand.* Auckland.
39. WARMKE, G., and ABBOTT, R. T., 1961. *Caribbean sea shells.* Narbeth, Pa.

Papers

40. ABBOTT, D. P. *et al.* (ed.). *Veliger*, 11, 1–109. (*Acmaea* biology).
41. ADAL, M. N., and MORTON, B. S., 1973. *J. Zool. Soc. Lond.,* 171, 533–556. (eye structure of *Laternula*).
42. ALLEN, J. A., 1954. *Quart. J. micr. Sci.,* 95, 473, 482. (*Pandora*).
43. ALLEN, J. A., 1958. *J. mar. Biol. Ass. U.K.,* 37, 97–112. (*Cochlodesma*).
44. ALLEN, J. A., 1958. *Phil. Trans. B.,* 241, 421–484. (Lucinacea).
45. ALLEN, J. A., 1968. *Proc. malacol. Soc. Lond.,* 38, 27. (form and function of Astartacea).

46. ALLEN, J. A., and SANDERS, H. L., 1969. *Malacologia*, **7,** 381–396. (early bivalve feeding).
47. ALLEN, J. A., 1975. *Proc. malacol. Soc. Lond.*, **41,** 601–609. (*Mesodesma*).
48. ANDREWS, E. B., 1964. *Proc. malacol. Soc. Lond.*, **35,** 121–40. (reproduction in Pilidae).
49. ANDREWS, Elizabeth B., 1964. *Proc. malacol. Soc. Lond.*, **36,** 121. (reproduction in pilid gastropods).
50. ANDREWS, Elizabeth B., and LITTLE, C., 1972. *J. Zool. Lond.*, **168,** 395–422. (excretion in Cyclophoridae).
51. ANKEL, W. E., 1937. *Biol. Zbl.*, **57,** 75–82. (boring in *Natica*).
52. ANSELL, A. D., 1967. *Proc. malacol. Soc. Lond.*, **37,** 395. (leaping in Asaphidae).
53. ANSELL, A. D., 1969. *Proc. malacol. Soc. Lond.*, **38,** 387. (leaping in Bivalvia).
54. ANSELL, A. D., and TRUEMAN, E. R., 1967. *J. exp. Biol.*, **46,** 105–116. (burrowing in Veneridae).
55. ATKINS, D., 1936–1943. *Quart. J. micr. Sci.*, **79–80** (pts. 1–7). (bivalve gills and mantle cavity).
56. BARBER, V. C., 1968. *Symp. Zool. Soc. Lond.*, **23,** 37–62. (cephalopod statocyst).
57. BEEDHAM, G. E., 1958. *Quart. J. Micr. Sci.*, **99,** 181–197. (mantle of Bivalvia).
58. BEEDHAM, G. E., and TRUEMAN, E. R., 1967. *J. Zool.*, **151,** 215–231 and 1968. *Ibid.*, **154,** 443–451. (mantle and shell: Aplacophora and Polyplacophora).
59. BEEMAN, R. D., 1970. *Veliger*, **13,** 1–31. (*Phyllaplysia* reproduction).
60. BERG, C. J., 1974. *Behaviour*, **51,** 274–322. (Strombid locomotion).
61. BERG, C. J., 1975. *Bull. Mar. Sci.*, **25** (3), 307–317. (*Strombus* and *Xenophora*).
62. BERRINGTON, A., and HUGHES, G. M., 1973. *Proc. malacol. Soc. Lond.*, **40,** 399. (locomotion in *Aplysia*).
63. BERRY, A. J., LIM, R., and KUMAR, A. S., 1973. *J. Zool. Lond.*, **170,** 189–200. (reproduction in *Nerita birmanica*).
64. BERRY, A. J., LOONG, S. C., and THUM, H. H., 1967. *Proc. malacol. Soc. Lond.*, **37,** 325. (genital systems of Ellobiidae).
65. BERRY, E. W., 1928. *Quart. Rev. Biol.*, **3,** 92–108. (adaptations of fossil cephalopods).

66. BERRY, S. S., 1920. *Biol. Bull. Woods Hole.* **38**, 141–195. (cephalopod light production).
67. BERRY, S. S., 1952. *Bull. Calif. Fish Game,* **38**, 183–188. (*Opisthoteuthis*).
68. BIDDER, A. M., 1950. *Quart. J. micr. Sci.,* **91**, 1–43. (digestion in Loliginidae).
69. BIDDER, A. M., 1957. *Pubbl. Staz. Zool. Napoli,* **29**, 139–15. (*Octopus digestion*).
70. BIDDER, A. M., 1962. *Nature,* **194**, 451–454. (*Nautilus*).
71. BIGHEN, A., and FARLEY, J., 1974. *Proc. malacol. Soc. Lond.,* **41**. (*Littorina* digestive gland cycles).
72. BILGIN, F. H., 1973. *Proc. malacol. Soc. Lond.,* **40**, 379. (form and function in *Melanopsis*).
73. BOETTGER, C. R., 1956. *Zool. Anz. Suppl.,* **19**, 223–256. (Aplacophora).
74. BOUILLON, J., 1960. *Ann. Sci. Nat. (Zool.),* **12** (2), 719–749. (ultrastructure of renal cells).
75. BOYCOTT, B. B., 1953. *Proc. Linn. Soc. Lond.,* **164**, 235–240. (chromatophore system of cephalopods).
76. BOYCOTT, B. B., 1960. *Proc. roy. Soc. B.,* **152**, 78–87. (*Octopus* statocysts).
77. BRADLEY, E. A., 1974. *J. Zool. Lond.,* **173**, 355–368. (habits of reared *Octopus*).
78. BROWN, S. C., 1969. *Malacologia,* **9** (2), 447–500. (*Nassarius* digestion).
79. CAIN, A. J., and SHEPPARD, P. M., 1954. *Genetics,* **39**, 89–116. (Natural selection in *Cepaea*).
80. CALOW, P., 1974. *Proc. malacol. Soc. Lond.,* **74**, 41. (*Planorbis* bacterial feeding).
81. CARRIKER, M. R., 1946. *Biol. Bull. Woods Hole,* **91**, 88–111. (digestion in *Lymnea*).
82. CARRIKER, M. R., 1959. *Proc. xv. Int. Cong. Zool. Lond.,* 373–6. (Muricid drilling).
83. CASTILLA, J. C., 1974. *Veliger,* **16** (3), 291–2. (*Concholepas* mating).
84. CHEESEMAN, D. F., 1956. *Nature,* **178**, 987. (feeding in *Ampullarius*).
85. CLARKE, M. R., 1966. *Adv. mar. Biol.,* **4**, 91–300. (oceanic squids).
86. CLARKE, M. R., 1970. *J. mar. Biol. Ass. U.K.,* **50**, 53–64. .(*Spirula spirula*).

87. CLELAND, Doreen M., 1954. *Proc. malacol. Soc. Lond.,* **30,** 167–202. (*Valvata*).
88. COMFORT, A., 1951. *Biol. Rev.,* **26,** 285–301. (shell pigments).
89. COMFORT, A., 1957. *Proc. malacol. Soc. Lond.,* **32,** 219–241. (age and growth).
90. COOK, P. M., 1949. *Proc. malacol. Soc. Lond.,* **27,** 265–271. (ciliary feeding in *Viviparus*).
91. COX, L. R., 1960. *Proc. malacol. Soc. Lond.,* **34,** 60–88. (bivalve classification).
92. CRAIG, A. K., *et al.,* 1969. *Amer. Zoologist,* **9,** 895–901. (*Siphonaria* and beach-rock destruction).
93. CRAMPTON, Denise M., 1975. *Proc. malac. Soc. Lond.,* **41,** 549–570. (*Testacella* feeding).
94. CREEK, Gwendoline A., 1953. *Proc. malacol. Soc. Lond.,* **29,** 228–240. (*Acme*).
95. CROFTS, Doris R., 1937. *Phil. Trans. B.,* **228,** 219–268. (*Haliotis* development).
96. CROFTS, Doris R., 1955. *Proc. zool. Soc. Lond.,* **125,** 711–750. (torsion).
97. DAINTON, Barbara H., 1954. *J. exp. Biol.,* **31,** 165–197. (activity of slugs).
98. DAVIS, J. D., 1968. *Proc. malacol. Soc. Lond.,* **38,** 135. (behaviour in scaphopod *Cadulus*).
99. DAY, Jennifer A., 1969. *Amer. Zoologist,* **9,** 909–916. (Cymatiid feeding).
100. DENTON, E. J., 1974. *Proc. Roy. Soc. Lond.* (B), **185,** 273–299. (cephalopod flotation and buoyancy).
101. DENTON, E. J., and GILPIN BROWN, J. B., 1961. *J. mar. Biol. Ass. U.K.,* **41,** 319–365. (buoyancy of *Sepia*).
102. DENTON, E. J., and GILPIN BROWN, J. B., 1966. *J. mar. Biol. Ass. U.K.,* **46,** 723–759. (buoyancy of *Nautilus*).
103. DENTON, E. J., and GILPIN BROWN, J. B., 1967. *J. mar. Biol. Ass. U.K.,* **47,** 723–759. (*Nautilus* buoyancy).
104. DENTON, E. J., GILPIN BROWN, J. B., and HOWARTH, J. V., 1967. *J. mar. Biol. Ass. U.K.,* **47,** 181–192. (*Spirula* buoyancy).
105. DINAMANI, P., 1963. *Proc. malacol. Soc. Lond.,* **36,** 1–5. (feeding in *Dentalium*).
106. DODD, J. N., 1956. *J. mar. Biol. Assoc. U.K.,* **35,** 327–340. (hermaphrodite *Patella*).

107. DONOVAN, D. T., 1964. *Biol. Rev.*, **39**, 259–287. (cephalopod classification).
108. DOUVILLÉ, H., 1912. *Bull. Soc. Geol. France* (4), **12**, 419–467. (bivalve classification).
109. DREW, G. A., 1907. *Biol. Bull. Woods Hole,* **12**, 127–138. (movement and habits in *Ensis*).
110. DREW, G. A., 1919. *J. Morph.*, **32**, 379–418. (spermatophore of *Loligo*).
111. DUNCAN, C. J., 1960. *Proc. zool. Soc. Lond.*, **134**, 601–609. (pulmonate genital system).
112. DUNCAN, C. J., 1960. *Proc. zool. Soc. Lond.*, **135**, 339–355. (genitalia in Basommatophora).
113. EALES, Nellie B., 1950. *Proc. malacol. Soc. Lond.*, **28**, 185–196. (secondary symmetry in gastropods).
114. EDMUNDS, M., 1966. *J. Linn. Soc.* (*Zool.*), **47**, 27–71. (protection in Aeolidacea).
115. EDMUNDS, M., 1966. *Proc. malacol. Soc. Lond.*, **37**, 73. (*Stiliger* defensive adaptations).
116. EDMUNDS, M., 1968. *Proc. malacol. Soc. Lond.*, **38**, 21. (Doridacea: acid secretion).
117. EDWARDS, D. C., 1969. *Amer. Zool.*, **9**, 399–417. (*Olivella*).
118. EVANS, F. G. C., 1951. *J. Anim. Ecol.*, **20**, 1–10. (habits of chitons).
119. EVANS, T. J., 1922. *Quart. J. micr. Sci.*, **66**, 439–455. (*Calma*).
120. EVANS, T. J., 1952. *Proc. malacol. Soc. Lond.*, **28**, 249–258. (*Alderia*).
121. FARMER, W. M., 1970. *Veliger*, **13**, 73. (swimming gastropods).
122. FLOWER, R. H., 1955. *J. Palaeont.*, **29**, 857–867. (fossil nautiloid trails).
123. FLOWER, R. H., 1961. *J. Palaeont.*, **35**, 569–74. (cephalopod classification).
124. FORREST, J. E., 1953. *Proc. Linn. Soc. Lond.*, **164**, 225–234. (dorid feeding).
125. FOSTER-SMITH, R. L., 1975. *Proc. malacol. Soc. Lond.*, **41**, 571–588. (mucus in bivalve feeding).
126. FRETTER, Vera, 1937. *Trans. roy. Soc. Edin.*, **59**, 119–164. (digestion in chitons).
127. FRETTER, Vera, 1939. *Proc. roy. Soc. Edin.*, **59**, 599–646. (digestion in tectibranchs).
128. FRETTER, Vera, 1943. *J. mar. Biol. Ass. U.K.*, **25**, 685–720. (*Oncidiella*).

129. FRETTER, Vera, 1946. *J. mar. Biol. Ass. U.K.*, **25**, 173–211; **26**, 213–351. (genital ducts in Gastropoda).

130. FRETTER, Vera, 1948. *J. mar. Biol. Ass. U.K.*, **27**, 597–632. (*Skeneopsis, Omalogyra, Rissoella*).

131. FRETTER, Vera, 1951. *J. mar. Biol. Ass. U.K.*, **29**, 567–586. (*Triphora* and *Cerithiopsis*).

132. FRETTER, Vera, 1955. *Proc. malacol. Soc. Lond.*, **31**, 137–144. (*Balcis*).

133. FRETTER, Vera, 1960. *Proc. zool. Soc. Lond.*, **135**, 537–549. (*Ringicula*).

134. FRETTER, Vera, 1965. *J. Zool.*, **147**, 46–74. (Neritacea).

135. FRETTER, Vera, 1967. *Proc. malacol. Soc. Lond.*, **37**, 357. (prosobranch veliger).

136. FRETTER, Vera, 1969. *Proc. malacol. Soc. Lond.*, **38**, 375. (metamorphosis in prosobranchs).

137. FRETTER, Vera, and GRAHAM, A., 1949. *J. mar. Biol. Ass. U.K.*, **28**, 493–532. (Pyramidellidae).

138. FRETTER, Vera, and GRAHAM, A., 1954. *J. mar. Biol. Ass. U.K.*, **33**, 565–585. (*Actaeon*).

139. FRETTER, Vera, and MONTGOMERY, M. C., 1968. *J. mar. Biol. Ass. U.K.*, **58**, 498–720. (gastropod veliger).

140. FRYER, G., 1961. *Phil. Trans. B.*, **244**, 259–298. (*Mutela*).

141. GAGE, J., 1966. *J. mar. Biol. Ass. U.K.*, **46**, 71–89. (*Montacuta*).

142. GALTSOFF, P. S., 1964. *U.S. Fish. and wildlife Bulletin*, **64**. (oyster biology).

143. GASCOIGNE, T., 1956. *Proc. roy. Soc. Edin.*, **63**, 129–151. (*Limapontia*).

144. GASCOIGNE, T., and SARTORY, P. K., 1974. *Proc. malacol. Soc. Lond.*, **41**, 109. (Sacoglossan dentition).

145. GHISELIN, M. T., 1965. *Malacologia*, **3**, 327–378. (opisthobranch genital system).

146. GILMOUR, T. H. J., 1962. *Proc. malacol. Soc. Lond.*, **35**, 81–85. (*Lima*).

147. GOREAU, T. F., and N. I., YONGE, C. M. and NEUMANN, Y., 1970. *J. Zool. Lond.*, **160**, 159–172. (*Fungicavia* feeding).

148. GOREAU, T. F., and N. I., and YONGE, C. M., 1973. *J. Zool. Soc. Lond.*, **169**, 417–454. (*Tridacna* zooxanthellae).

149. GRAHAM, A., 1931. *Trans. roy. Soc. Edin.*, **56**, 725–751. (*Ensis*).

150. GRAHAM, A., 1932. *Trans. roy. Soc. Edin.*, **57**, 287–308. (digestion in limpet).

151. GRAHAM, A., 1934. *Proc. roy. Soc. Edin.*, **54**, 158–187. (Tellinacea).
152. GRAHAM, A., 1938. *Proc. zool. Soc. Lond.*, **108** (A), 453–463. (*Turritella*).
153. GRAHAM, A., 1938. *Proc. roy. Soc. Edin.*, **59**, 267–307. (gut in aeolids).
154. GRAHAM, A., 1939. *Proc. zool. Soc. Lond.*, **109** (B), 75–112. (style-bearing prosobranchs).
155. GRAHAM, A., 1949. *Trans. roy. Soc. Edin.*, **41**, 737–778. (molluscan stomach).
156. GRAHAM, A., 1955. *Proc. malacol. Soc. Lond.*, **31**, 144–159. (origins, feeding).
157. GRAHAM, A., 1964. *Proc. zool. Soc. Lond.*, **143**, 301–329. (*Patella* buccal mass).
158. GRAHAM, A., 1965. *Proc. malacol. Soc. Lond.*, **34**, 323–338. (*Ianthina*).
159. GREENAWAY, P., 1970. *J. exp. Biol.*, **53**, 147–163. (*Lymnaea* regulation).
160. GREENE, R. W., 1970. *Mar. Biol. Wash.*, **7**, 138–142. (symbiotic chloroplasts in *Sacoglossa*).
161. GREENFIELD, L. J., and LANE, C. E. *J. biol. Chem.*, **204**, 669–672. (cellulose digestion in *Teredo*).
162. GREENFIELD, M. L., 1972. *J. Zool. Lond.*, **166**, 37–47. (*Acanthopleura* feeding and digestion).
163. HADFIELD, M. G., 1969. *Veliger*, **12** (3), 301–309. (*Petaloconchus* and *Serpulorbis*).
164. HEMBROW, Deborah, 1973. *Proc. malacol. Soc. Lond.*, **40**, 505. (*Planorbarius* buccal mass).
165. HOLME, N. A., 1961. *J. mar. Biol. Ass. U.K.*, **41**, 699–703. (Tellinacea).
166. HOLMES, W., 1949. *Proc. zool. Soc. Lond.*, **110**, 17–36. (*Sepia* colour change).
167. HOUBRICK, J. R., and FRETTER, Vera, 1969. *Proc. malacol. Soc. Lond.*, **38**, 415. (form and function in *Cymatium* and *Bursa*).
168. HOWELLS, H. H., 1942. *Quart. J. micr. Sci.*, **83**, 357–397. (gut of *Aplysia*).
169. HUBBARD, S. J., 1960. *J. exp. Biol.*, **37**, 845–853. (*Octopus* statocysts and hearing).
170. HUGHES, R. N., and LEWIS, A. H., 1974. *J. Zool. Lond.*, **172**, 531–547. (*Dendropoma*: feeding and reproduction).

171. HUNTER, W. R., 1955. *Studies on Loch Lomond,* I. Univ. Glasgow. (freshwater pulmonates).
172. HURST, A., 1965. *Malacologia,* **2**, 281–347. (feeding in *Philine*).
173. JACCARINI, V., BANNISTER, W. H., and MICALLEF, H., 1967. *J. Zool. Lond.,* **154**, 397–410. (*Lithophaga* boring).
174. JONES, E. C., 1963. *Science,* **139**, 764–766. (*Tremoctopus* and *Physalia*).
175. JØRGENSON, C. B., 1955. *Biol. Rev.,* **30**, 391–454. (ciliary feeding).
176. JUDD, W., 1971. *Proc. malacol. Soc. Lond.,* **39**, 343. (form and function in galeommatid *Divariscintilla*).
177. KASINATHAN, R., 1975. *Proc. malacol. Soc. Lond.,* **41**, 379–394. (Cyclophoridae).
178. KAWAGUTI, S., 1950. *Pacific Science,* **4**, 43–49. (*Corculum* and zooxanthellae).
179. KAWAGUTI, S., and BABA, K., 1959. *Biol. J. Okayama Univ.,* **5**, 117–184. (*Tamanovalva*).
180. KAY, E. Alison, 1968. *Symp. Zool. Soc. Lond.,* **22**, 109–134. (bivalved Sacoglossa).
181. KELLOGG, J. L., 1915. *J. Morph.,* **26**, 625–701. (bivalve mantle cavities).
182. KNIGHT, J. B., 1952. *Smithson misc. Coll.,* **117**, no. 13. (primitive fossil gastropods).
183. KNIGHT, J. B., and YOCHELSON, E. L., 1958. *Proc. malacol. Soc. Lond.,* **33**, 37–48. (Monoplacophora and early gastropods).
184. KNIGHT, J. B., and YOCHELSON, E, L., 1960. In *Treatise on Invertebrate Palaeontology* (R. C. Moore, ed.), vol. 1, 77–84. (Monoplacophora: Palaeacmaeidae).
185. KOHN, A. J., 1956. *Proc. Nat. Acad. Sci.,* **42**, 168–171. (feeding in *Conus*).
186. KOHN, A. J., 1961. *Amer. Zoologist,* **1**, 291–308. (Chemoreception in piscivorous molluscs).
187. KORRINGA, P., 1952. *Quart. Rev. Biol.,* **27**, 266–308; 339–365. (oyster biology).
188. LEMCHE, H., 1957. *Nature,* **179**, 413–416. (*Neopilina*).
189. LEMCHE, H., and WINGSTRAND, K. G., 1959. *Galathea Repts.,* **3**, 9–72. (*Neopilina*).
190. LILLY, Molly M., 1963. *Proc. malacol. Soc. Lond.,* **30**, 87–109. (*Bithynia*).

191. LIM, C. F., 1969. *Veliger,* **12,** 160–4. (feeding of *Conus*).
192. LIND, H., 1973. *J. Zool. Lond.,* **169,** 39–64. (*Helix*: functions of spermatophore).
193. LINSLEY, R. M., 1978. *Amer. Scientist,* **66,** 432–441. (gastropod shell form and evolution).
194. LISSMANN, H. W. 1945. *J. exp. Biol.,* **21,** 58–69. (gastropod movement).
195. LITTLE, C., 1967. *J. exp. Biol.,* **46,** 459–474. (*Strombus* excretion).
196. LITTLE, C., 1972. *J. exp. Biol.,* **56,** 249–261. (excretion in land Neritacea).
197. LUTZEN, J., and NIELSEN, K., 1975. *Vid. Medd. Dansk. Naturhist. For.,* **138,** 171–199. (*Echineulima,* a parasitic eulimid).
198. MCGINITIE, G., 1941. *Biol. Bull. Woods Hole,* **80,** 18–25. (bivalve food collecting).
199. MACHIN, J., 1966. *J. exp. Biol.,* **45,** 269–278. (*Helix* water loss).
200. MCLEAN, J. H., 1962. *Proc. malacol. Soc. Lond.,* **35,** 23–26. (*Placiphorella*).
201. MCQUISTON, R. W., 1969. *Proc. malacol. Soc. Lond.,* **38,** 483–492. (*Lasaea* digestive diverticula).
202. MARSH, Helene, 1971. *Veliger,* **14** (1), 45–53. (Vermivorous *Conus*).
203. MARTIN, A. W., and HARRISON, F. M. In *Physiology of Mollusca* (K. M. Wilbur and C. M. Yonge, eds.), 353–386. (excretion).
204. MARTIN, R., 1966. *Pubbl. Staz. Zool. Napoli,* **35,** 61–75. (swimming of *Notarchus*).
205. MATHERS, N. F., 1974. *Proc. malacol. Soc. Lond.,* **41,** 89. (filter feeding in oysters).
206. MILLER, M. C., 1961. *J. Anim. Ecol.,* **30,** 95–116. (Nudibranch feeding).
207. MILLER, M. C., 1962. *J. Anim. Ecol.,* **31,** 545–569. (Nudibranch annual cycles).
208. MILLER, Susanne L., 1974. *Proc. malacol. Soc. Lond.,* **41,** 233. (prosobranch locomotion).
209. MILLOTT, N., 1937. *Phil. Trans. B.,* **228,** 173–217. (feeding in dorid *Jorunna*).
210. MOORE, H. J., 1971. *Marine Biology,* **11,** 22–27. (laterofrontal cirri).

211. MORTON, B. S., 1969. *Proc. malacol. Soc. Lond.*, **38**, 301 and 401. (*Dreissena*: form and function – rhythms).
212. MORTON, B. S., 1970. *Biol. J. Linn. Soc.*, **3**, 329–342. (*Ostrea*: feeding and digestive rhythms).
213. MORTON, B. S., 1970. *Proc. malacol. Soc. Lond.*, **39**, 151–167. (*Teredo* feeding and digestion).
214. MORTON, B. S., 1973. *Malacologia*, **14**, 63–79. (bivalve feeding and digestion).
215. MORTON, B. S., and MCQUISTON, R. W., 1974. *Forma et functio*, **7**, 59–80. (*Teredo* rhythms).
216. MORTON, J. E., 1951. *Quart. J. micr. Sci.*, **92**, 1–20. (*Struthiolaria*).
217. MORTON, J. E., 1951. *Proc. roy. Soc. N.Z.*, **79**, 1–51. (Vermetidae).
218. MORTON, J. E., 1954. *Discovery Repts.*, 163–200. (sex in *Limacina*).
219. MORTON, J. E., 1954. *J. mar. Biol. Ass. U.K.*, **33**, 297–312. (*Limacina*).
220. MORTON, J. E., 1954. *Proc. zool. Soc. Lond.*, **125**, 127–168. (pulmonate evolution).
221. MORTON, J. E., 1955. *J. mar. Biol. Ass. U.K.*, **34**, 113–149. (*Otina*).
222. MORTON, J. E., 1955. *Phil. Trans. B.*, **239**, 89–160. (Ellobiidae).
223. MORTON, J. E., 1956. *J. mar. Biol. Ass. U.K.*, **35**, 563–586. (digestion in *Lasaea*).
224. MORTON, J. E., 1958. *J. mar. Biol. Ass. U.K.*, **37**, 287–97. (*Clione*).
225. MORTON, J. E., 1958. *Proc. malacol. Soc. Lond.*, **34**, 1–10. (torsion).
226. MORTON, J. E., 1959. *J. mar. Biol. Ass. U.K.*, **38**, 225–238. (*Dentalium*).
227. MORTON, J. E., 1960. *Biol. Rev.*, **33**, 92–140. (ciliary feeding gut).
228. MORTON, J. E., 1960. *J. mar. Biol. Ass. U.K.*, **39**, 5–26. (orientation of *Lasaea*).
229. MORTON, J. E., 1963. *Proc. Linn. Soc.*, **174**, 53–72. (molluscan classification).
230. MORTON, J. E., 1965. *Bull. Brit. Mus. N.H.*, **11**, 583–630. (Vermetidae).
231. MORTON, J. E., and HOLME, N. A., 1955. *J. mar. Biol. Ass. U.K.*, **34**, 101–112. (*Akera*).

232. NAIR, N. B., and ANSELL, A. D., 1968. *Proc. malacol. Soc. Lond.*, **38**, 179. (bivalve burrowing).
233. NARCHI, W., 1975. *Proc. malac. Soc. Lond.*, **41**, 451–465. (*Petricola*: functional morphology).
234. NEEDHAM, J., 1938. *Biol, Rev.*, **13**, 225–251, (uricotely in gastropods).
235. NEWELL, G, E., 1953. *J. mar. Biol. Ass. U.K.*, **37**, 296–266. (orientation in *Littorina*).
236. NEWELL, G. E., 1958. *J. mar. biol. Ass. U.K.*, **37**, 229–239; 214–266. (orientation in *Littorina*).
237. NEWELL, N. D., 1969. In *Treatise on Invertebrate Palaentology* (R. C. Moore, ed.). (bivalve classification).
238. NEWELL, R. C., 1962. *Proc. zool. Soc. Lond.*, **138**, 49–75. (behaviour of *Hydrobia*).
239. NISBET, R. H., 1973. *Soc. Lond.*, **40** (6), 435–468. (*Trochus* buccal mass).
240. ODHNER, N. H., 1939. *Kgl. Norske Vid. Selske. Skr.*, **1**, 1–93. (opisthobranch classification).
241. OLDFIELD, Eileen, 1955. *Proc. malacol. Soc. Lond.*, **31**, 226–249. (*Lasaea* and *Turtonia*).
242. OLDFIELD, Eileen, 1964. *Proc. malacol. Soc. Lond.*, **36**, 79–120. (reproduction of Montacutidae).
243. ORTON, J. H., 1912. *J. mar. Biol. Ass. U.K.*, **9**, 444–478. (feeding in *Crepidula*).
244. ORTON, J. H., SOUTHWARD, A. J., and DODD, J. M., 1956. *J. mar. Biol. Ass. U.K.*, **35**, 149–176. (sex in limpets).
245. OWEN, G., 1953. *J. mar. Biol. Ass. U.K.*, **32**, 85–105. (*Isocardia*).
246. OWEN, G., 1953. *Quart. J. micr. Sci.*, **94**, 57–70. (bivalve shell form).
247. OWEN, G., 1955. *Quart. J. micr. Sci.*, **96**, 517–537. (bivalve digestive gland).
248. OWEN, G., 1956. *Quart. J. micr. Sci.*, **97**, 541–568. (gut of *Nucula*).
249. OWEN, G., 1966. In *Physiology of Mollusca*, vol. 2. N.Y.: Academic Press. (digestion).
250. OWEN, G., 1970. *Phil. Trans. B.*, **258**, 245–260. (*Cardium* digestive gland).
251. OWEN, G., TRUEMAN, E. R., and YONGE, C. M., 1952. *Nature*, **171**, 73–75. (ligament in bivalves).

252. PACKARD, A., 1972. *Biol. Rev.*, **47**, 241–307. (cephalopods and fish).
253. PAL, S. G., 1972. *Proc. malacol. Soc. Lond.*, **40**, 161. (*Mya*: digestive cells).
254. PICKEN, L. E. R., 1937. *J. exp. Biol.*, **14**, 20–34. (excretion in *Anodonta* and *Lymnea*).
255. PICKFORD, Grace, 1946. *Dana Repts.*, **29**, 40 pp. (*Vampyroteuthis*).
256. POPHAM, M. L., 1940. *J. mar. Biol. Ass. U.K.*, **24**, 549–587. (Erycinacea).
257. POTTS, F. A., 1923. *Biol. Rev.*, **1**, 1–16. (*Teredo*).
258. POTTS, W. T. W., 1954. *J. exp. Biol.*, **31**, 613–617. (osmoregulation in bivalves).
259. POTTS, W. T. W., 1967. *Biol. Rev.*, **42**, 1–41. (excretion).
260. PURCHON, R. D., 1941. *J. mar. Biol. Ass. U.K.*, **25**, 1–39 (*Xylophaga*).
261. PURCHON, R. D., 1955. *Proc. zool. Soc. Lond.*, **124**, 859–911. (Pholadidae).
262. PURCHON, R. D., 1956. *Proc. zool. Soc. Lond.*, **124**, 245–258. (*Martesia*).
263. PURCHON, R. D., 1956–60. *Proc. zool. Soc. Lond.*, **127**, 511–525; **129**, 27–60; **135**, 431–489. (bivalve stomachs).
264. PURCHON, R. D., 1960. *Proc. malacol. Soc. Lond.*, **33**, 224–230. (bivalve classification).
265. PURCHON, R. D., 1962. *Proc. malacol. Soc. Lond.*, **62**, 35. (bivalve classification).
266. PURCHON, R. D., 1963. *Proc. malacol. Soc. Lond.*, **35**, 251–272. (*Egeria*).
267. PURCHON, R. D., 1971. *Proc. malacol. Soc. Lond.*, **39**, 253–262. (digestion in bivalves).
268. QUAYLE, D. B., 1949. *Proc. malacol. Soc. Lond.*, **28**, 31–37. (burrowing of *Venerupis*).
269. RAO, M. B., 1975. *Proc. malacol. Soc. Lond.*, **41**, 309. (digestion in limpets, *Cellana*).
270. REES, W. J., 1964. *Proc. malacol. Soc. Lond.*, **36**, 55. (breathing devices in land operculates).
271. REID, Jocelyn D., 1964. *Proc. zool. Soc. Lond.*, **143**, 365–393. (*Elysia*).
272. REID, R. G. B., and REID, M. Alison, 1974. *Sarsia*, **56**, 47–56. (*Cuspidaria*).

273. ROBERTSON, J. D., 1953. *J. exp. Biol.*, **30**, 277–296. (molluscan excretion).
274. ROBERTSON, R., 1963. *Proc. malacol. Soc. Lond.*, **35**, 51–64. (Epitoniidae).
275. ROBINSON, Elizabeth, 1960. *Proc. zool. Soc. Lond.*, **135**, 319–338. (Toxoglossa).
276. SALEUDDIN, A. S. M., 1964. *Proc. malacol. Soc. Lond.*, **36**, 149–162. (*Cyprina*).
277. SALEUDDIN, A. S. M., 1965. *Proc. malacol. Soc. Lond.*, **36**, 229–258. (*Astarte*).
278. SALVINI-PLAWEN, L., 1969. *Malacologia*, **9** (1), 191–216. (Aplacophora: organization and phylogeny).
279. SOLIMAN, G. N., 1973. *Proc. malacol. Soc. Lond.*, **40**, 313. (form and function in *Rocellaria*).
280. SPATHE, L. F., 1933. *Biol. Rev.*, **8**, 418–462. (evolution of fossil Cephalopoda).
281. STASEK, C. R., 1961. *Proc. zool. Soc. Lond.*, **137**, 511–538. (early bivalve feeding).
282. STASEK, C. R., 1965. *J. Morph.*, **112**, 195–214. (bivalve form).
283. STASEK, C. R., 1972. Ch. 1 in *Chemical Zoology*: N.Y. and Lond.: Academic Press. (molluscan origins and phylogeny).
284. SUMNER, A. T., 1966. *J. roy. micr. Soc.*, **85**, 417–423. (*Anodonta*: fine structure of digestive cells).
285. SUTHERLAND, N. S., 1960. *J. comp. physiol. Psychol.*, **53**, 104–112. (*Octopus*: shape discrimination).
286. SUTHERLAND, N. S., 1961. *J. comp. physiol. Psychol.*, **54**, 43–48. (*Octopus*: dimensional discrimination).
287. TAYLOR, D. L., 1968. *J. mar. Biol. Ass. U.K.*, **48**, 1–15. (*Elysia*: chloroplasts and symbiosis).
288. THOMAS, R. F., 1973. *Proc. malacol. Soc. Lond.*, **40**, 303. (*Siphonaria*: homing and rhythms).
289. THOMPSON, T. E., 1960. *Proc. roy. Soc. B.*, **153**, 263–278. (*Neomenia*).
290. THOMPSON, T. E., 1961. *Quart. J. micr. Sci.*, **102**, 1–14. (*Tritonia* genital system).
291. THOMPSON, T. E., 1962. *Phil. Trans. B.*, **245**, 172–218. (development of *Tritonia*).
292. THOMPSON, T. E., 1967. *J. mar. Biol. Ass. U.K.*, **47**, 1–22. (nudibranch development).
293. THOMPSON, T. E., and SLINN, D. J., 1959. *J. mar. Biol. Ass. U.K.*, **38**, 507–524. (*Oscanius*).

294. THORPE, N., 1972. *J. exp. Biol. Ecol.* (rhythms of *Macoma* and *Scrobicularia*).
295. TRUEMAN, E. R., 1951. *Quart. J. micr. Sci.*, **92**, 129–140. (bivalve hinge and ligament).
296. TRUEMAN, E. R., 1968. *Proc. malacol. Soc. Lond.*, **38**, 139. (*Mactra* burrowing).
297. TRUEMAN, E. R., 1971. *J. Zool. Soc. Lond.*, **165**, 453–469. (burrowing and migration in *Donax*).
298. TRUEMAN, E. R., BRAND, A. R., and DAVIS, P., 1966. *J. exp. Biol.*, **44**, 469–492. (bivalve burrowing).
299. ULBRICK, Martha L., 1969. *Proc. malacol. Soc. Lond.*, **38**, 431. (*Crucibulum spinosum*).
300. VERMEIJ, G. J., 1971. *J. Zool. Lond.*, **163**, 15–23. (shell geometry and evolution in gastropods).
301. WAGGE, L., 1952. *Quart. J. micr. Sci.*, **92**, 307–322. (shell repair in *Helix*).
302. WALSBY, J., MORTON, J. E., and CROXALL, J. P., 1973. *J. Zool. Lond.*, **171** (2), 257–283. (Gadinalia mucus-feeding).
303. WEBBER, H. H., and DEHNEL, P. A., 1968. *Comp. Biochem. Physiol.*, **25**, 49–64. (*Acmaea* ionic regulation).
304. WELLS, M. J., 1959. *J. exp. Biol.*, **36**, 590–612. (tactile learning in *Octopus*).
305. WELLS, M. J., 1961. *J. exp. Biol.*, **38**, 127–133. (*Octopus*: weight discrimination).
306. WELLS, M. J., 1961. *Advanc. Sci. London*, **20**, 461–471. (senses of *Octopus*).
307. WELLS, M. J., 1963. *J. exp. Biol.*, **40**, 187–193. (*Octopus*: taste and touch).
308. WELLS, M. J., 1964. *J. exp. Biol.*, **41**, 433–445. (*Octopus*: shape and discrimination).
309. WELLS, M. J., and WELLS, J., 1957. *J. exp. Biol.*, **34**, 131–142. (tactile discrimination in *Octopus*).
310. WELLS, M. J., and WELLS, J., 1960. *J. exp. Biol.*, **37**, 489–499; (*Octopus*: proprioception and vision).
311. WELLS, M. J., and WELLS, J., 1961. *J. exp. Biol.*, **38**, 811–826. (*Octopus*: tactile and visual learning).
312. WERNER, B., 1953. *Zool. Anz. Suppl.*, **17**, 529–546. (ciliary feeding in Prosobranchia).
313. WILSON, B. R., 1968. *Veliger*, **38**, 121. (swimming propodium in Olividae).

313. YOCHELSON, E. L., FLOWER, R. H., and WEBERS, G. F., 1973. *Lethaia*, **6**, 275–310. (monoplacophoran origins of cephalopods).
314. YOLOYE, V., 1975. *Proc. malacol. Soc. Lond.*, **41**, 277. (form and function of *Anadara*).
315. YONGE, C. M., 1926. *J. Linn. Soc. Zool.*, **36**, 417. (feeding in Thecosomata).
316. YONGE, C. M., 1926. *J. mar. Biol. Ass. U.K.*, **14**, 295–386. (digestion in *Ostrea*).
317. YONGE, C. M., 1926. *Trans. roy. Soc. Edin.*, **54**, 703–718. (digestive diverticula).
318. YONGE, C. M., 1927. *Phil. Trans. B.*, **216**, 221–263. (Septibranchia).
319. YONGE, C. M., 1928. *Biol. Rev.*, **3**, 21–76. (feeding mechanisms).
320. YONGE, C. M., 1937. *Biol. Rev.*, **12**, 87–115. (digestive systems).
321. YONGE, C. M., 1937. *J. mar. Biol. Ass. U.K.*, **21**, 687–703. (*Aporrhais*).
322. YONGE, C. M., 1938. *J. mar. Biol. Ass. U.K.*, **22**, 453–468. (ciliary feeding prosobranchs).
323. YONGE, C. M., 1939. *Phil. Trans. B.*, **230**, 79–147. (Protobranchia).
324. YONGE, C. M., 1939. *Quart. J. micr. Sci.*, **81**, 367–390. (mantle cavity of chitons).
325. YONGE, C. M., 1946. *J. mar. Biol. Ass. U.K.*, **26**, 358–376. (*Aloidis*).
326. YONGE, C. M., 1947. *Phil. Trans. B.*, **232**, 442–518. (evolution of mantle cavity).
327. YONGE, C. M., 1949. *Phil. Trans. B.*, **234**, 29–76. (Tellinacea).
328. YONGE, C. M., 1953. *Proc. zool. Soc. Lond.*, **123**, 551–561. (Tridacnidae).
329. YONGE, C. M., 1953. *Trans. roy. Soc. Edin.*, **62**, 443–478. (monomyarian lamellibranchs).
330. YONGE, C. M., 1955. *Phil. Trans. B.*, **237**, 335–374. (*Pinna* and Aviculacea).
331. YONGE, C. M., 1955. *Quart. J. micr. Sci.*, **96**, 383–410. (boring in Mytilidae).
332. YONGE, C. M., 1957. *Nature*, **180**, 765–766. (*Aenigmonia*).
333. YONGE, C. M., 1957. *Pubbl. Staz. Zool. Mar. Napoli*, **29**, 151–170. (bivalve mantle and siphons).

334. YONGE, C. M., 1958. *Proc. malacol. Soc. Lond.,* **33,** 25–31. (*Petricola*).
335. YONGE, C. M., 1962. *J. mar. Biol. Ass. U.K.,* **42,** 113–125. (bivalve byssus and evolution).
336. YONGE, C. M., 1962. *Phil. Trans. B.,* **245,** 423–458. (evolution in Etheriidae).
337. YONGE, C. M., 1967. *Phil. Trans. B.,* **259,** 49–105. (evolution of Chamacea).
338. YONGE, C. M., 1967. *Proc. malacol. Soc. Lond.,* **37:** 311 (*Pedum*).
339. YONGE, C. M., 1969. *Proc. malacol. Soc., Lond.,* **38,** 493–527. (Carditacea).
340. YONGE, C. M., 1971. *Malacologia,* **11** (1), 1–44. (radiation of Saxicavacea).
341. YONGE, C. M., and NICHOLAS, H. M., 1940. *Bgk. Pap. Tortugas Lab.,* **32,** 289–301. (zooxanthellae in *Tridachia*).
342. YOUNG, D. K., 1969. *Amer. Zool.,* **9,** 903–907. (*Okadaia*: feeding).
343. YOUNG, D. K., 1969. *Malacologia,* **9** (2), 422–446. (dorid feeding).
344. YOUNG, J. Z., 1939. *Phil. Trans. B.,* **229,** 465–503. (giant fibres).
345. YOUNG, J. Z., 1951. *Proc. roy. Soc. B.,* **139,** 18–37. (*Octopus*: learning).
346. YOUNG, J. Z., 1959. *Proc. Roy. Inst. G.B.,* **57,** 394–411. (cephalopods).
347. YOUNG, J. Z., 1959. *Proc. zool. Soc. Lond.,* **133,** 471–479. (*Argonauta*).
348. YOUNG, J. Z., 1960. *Nature* (*Lond.*), **186,** 836–839. (*Octopus*: visual system).
349. YOUNG, J. Z., 1960. *Proc. roy. Soc. B.,* **152,** 3–29. (*Octopus*: statocyst).
350. YOUNG, J. Z., 1961. *Biol. Rev.,* **36,** 32–96. (*Octopus*: learning).
351. YOUNG, J. Z., 1965. *Phil. Trans. B.,* **249,** 1–27. (*Nautilus*: brain).
352. YOUNG, J. Z., 1965. *Phil. Trans. B.,* **249,** 27–44; 54–67. (*Octopus*: brain).

Index

Page numbers in italics refer to line-drawings